SALT AFFECTED SOILS IN EUROPE

I. SZABOLCS

Chairman of the Subcommission on Salt Affected Soils
of the International Society of Soil Science

Director of the Research Institute for Soil Science
and Agricultural Chemistry of the Hungarian Academy
of Sciences

Springer-Science+Business Media, B.V.
1974

Additional material to this book can be downloaded from http://extras.springer.com

ISBN 978-94-011-8638-4   ISBN 978-94-011-9422-8 (eBook)
DOI 10.1007/978-94-011-9422-8

Contributors

Austria:	J. Fink
	O. Nestroy
Bulgaria:	L.P. Raikov
	H. Trashliev
Czechoslovakia:	J. Hrasko
France:	E. Servat
	J. Servant
Greece:	E.P. Papanicolaou
Hungary:	I. Szabolcs
	G. Várallyay
	J. Mélyvölgyi
Italy·	O.T. Rotini
	L. Carloni
Portugal:	J. Carvalho Cardoso
	M. Branco Marado
Rumania:	N. Florea
	J. Munteanu
Spain:	P. Grajera Torrez
	R. Grande Covian
	J. Bardaji Cando
	J. Manuel Ontanon
USSR:	V.V. Egorov
	N.I. Bazilevich
	E.I. Pankova
Yugoslavia:	N. Miljkovic
	G. Filipovski
	N. Plamenac
	P. Blaskovic

Contents

3

Preface

During the last decade it has been increasingly realized that the heavy demand on the limited land resources of the world has made it necessary to protect cultivated soils from degradation and to reclaim those areas which have lost their productivity. With special reference to saline and alkaline soils the International Society of Soil Science set up, in 1964, a Subcommission on Salt Affected Soils with a view to identifying and locating the salt affected soils of the world. This Subcommission was entrusted with the preparation of a world map of salt affected soils at scale 1:5,000,000, a project which it is conducting in cooperation with the Hungarian Academy of Sciences and Unesco.

The present publication covers the European section of the world map of salt affected soils. It is the outcome of an international cooperation of soil scientists from different European countries who contributed the material necessary for the compilation of the map. Cooperation was also established with the FAO/Unesco Soil Map of the World project in order to secure correlation with the international legend.

The profile descriptions included in this explanatory text and the criteria proposed for the subdivision of salt affected soils provide a basis for the transfer of information and experience within the framework of a global inventory of soil resources.

The publication draws attention to the potential hazards of salinity and alkalinity which may be a cause of failure of costly irrigation and land reclamation projects. These data should be made full use of by planners in their decisions for the utilization and improvement of land resources.

The Map of European Salt Affected Soils represents a first inventory of these soils in this region. It is a valuable contribution to an international approach aiming at a more effective use of the world's soil resources.

Rome - Moscow, June 1974

R. Dudal
Chairman, Commission V
International Society of Soil Science

V.A. Kovda
President
International Society of Soil Science

Introduction

This publication has been compiled within the framework of the programme for the preparation of the World Map of Salt Affected Soils. The Subcommission on Salt Affected Soils of the International Society of Soil Science was requested to accept responsibility for this project and has been supported in its activity by the Hungarian Academy of Sciences and by UNESCO from the very beginning.

The first meeting of the group of experts, discussing the possibilities as well as the ways and means of the construction of a map showing the worldwide extent and distribution of salt affected soils, was convened conjointly by UNESCO and the Board of the Subcommission in Budapest, October 2-5, 1967.

The Meeting was attended by the members of the Board: G. Aubert (France), C.A. Bower (USA), V.V. Egorov (USSR), M.M. Elgabaly (UAR), S.V. Govinda Rajan (India), V.A. Kovda (USSR), J.K. Skene (Australia), I. Szabolcs (Hungary) Chairman, as well as by F.A. van Baren (ISSS), S. Evteev (UNESCO) and J.H.V. van Baren (FAO).

The experts unanimously agreed that the preparation of such a map was of pressing importance because the salinity and alkalinity of soils hinder the satisfactory agricultural utilization of lands in many regions, particularly in irrigated areas. The better knowledge of the distribution of the various types of salt affected soils within a country makes the proper planning of agricultural production possible. The grouping of salt affected soils according to uniform principles allows their comparison and, consequently, all the experiences gained in the field of the reclamation and utilization of such soils may be shared to the mutual benefit of the countries concerned.

The experts agreed that the map should be a logical extension of the FAO/UNESCO Soil Map of the World Project, emphasizing the specific problems of salt affected soils for their better understanding on a global scale. The scale selected was 1 : 5,000,000, corresponding to that of the Soil Map of the World.

In order to promote the implementation of the project, the following regional coordinators were nominated:

Region	Co-ordinator
Africa	G. Aubert (France)
Australia	J.K. Skene (Australia)
Europe	I. Szabolcs (Hungary)
Middle East	M. Elgabaly (UAR)
North and Central America	C. Bower (USA)
South America	-[*]
South East Asia	S.V. Govinda Rajan (India)
USSR, Central Asia and the Far East	V.V. Egorov (USSR)

([*] At that time no co-ordinator could be nominated for South America and the map was to be prepared on the basis of a collection of relevant data available in the World Soil Resources Office of FAO, Rome.)

7

The Research Institute for Soil Science and Agricultural Chemistry of the Hungarian Academy of Sciences was appointed to serve as a centre for co-ordinating the World Map of Salt Affected Soils Project under the direction of I. Szabolcs.

Since the start of the programme up to the publication of the present volume several changes have taken place in connection with the co-ordinators:

a) A working group was formed to undertake the preparation of the map of salt affected soils for South America. Its members are: A. Zavaleta (Peru), R. Dudal (FAO) and I. Szabolcs (Hungary).

b) S.V. Govinda Rajan was replaced by J.S. Kanwar (India)

c) C. Bower retired and was replaced by J.D. Rhoades (USA)

d) J.K. Skene retired and was replaced by K.H. Northcote (Australia)

Australia was the first continent for which the map, together with an explanatory booklet, was completed and published by K.H. Northcote and J.K. Skene. It is entitled: "Australian Soils with Saline and Sodic Properties" and edited as SOIL PUBLICATION No. 27. COMMONWEALTH SCIENTIFIC AND INDUSTRIAL RESEARCH ORGANISATION, AUSTRALIA, 1972.

For the preparation of the Map of European Salt Affected Soils a working group was formed. Its first meeting was held in Novi Sad, Yugoslavia, on May 21-24, 1968. The participants deliberated on the problems of the grouping and classification of European salt affected soils.

Problems encountered in the course of the preparation of the maps and new suggestions were discussed at a special session during the 9th Congress of I.S.S.S. in Adelaide, Australia, in August 1968, as well as during the next meeting of the European working group in Budapest, Hungary, in December 1968.

The first draft of the Map of European Salt Affected Soils was presented during the Symposium of the Subcommission on Salt Affected Soils of the I.S.S.S. in Yerevan, Armenian SSR, in May 1969. The second, revised map was discussed by the working group in Smolenice, Czechoslovakia, in February 1970 and was accepted in its final form during the next meeting of the Subcommission in Sevilla, Spain, in May 1971.

The experts, who have contributed to the compilation of the Map of European Salt Affected Soils, are as follows:

Australia:	J. Fink
	O. Nestroy
Bulgaria:	L.P. Raikov
	H. Trashliev
Czechoslovakia:	J. Hrasko
France:	E. Servat
	J. Servant
Greece:	E.P. Papanicolaou
Hungary:	I. Szabolcs
	G. Várallyay
	J. Mélyvölgyi
Italy:	O.T. Rotini
	L. Carloni

Portugal:	J. Carvalho Cardoso
	M. Branco Marado
Rumania:	N. Florea
	J. Munteanu
Spain:	P. Grajera Torrez
	R. Grande Covian
	J. Bardaji Cando
	J. Manuel Ontanon
USSR[*]	V.V. Egorov
	N.I. Bazilevich
	E.I. Pankova
Yugoslavia:	N. Miljkovic
	G. Filipovski
	N. Plamenac
	P. Blaskovic

* Parts of the material were presented by: S.M. Ahnoyan, G.D. Chernoyva, V.I. Chikvishvili, N.I. Gorina, G.S. Grin, - F.I. Kozlovsky, E.M. Mirzoev, A.B. Novikova, K.A. Oganesian, S.S. Piruzian, N.N. Rozov, V.R.V obuev, B.A. Zimovetz, L.T. Zveryova.

Definition and Grouping of Salt Affected Soils

It is generally accepted that water soluble salts, particularly the sodium salts are responsible for the low fertility of salt affected soils. Saline or alkali soils are soils of which the content of salts (or their ions) interferes with the growth of the majority of crops.

Two main groups of these soils may be distinguished:

1. Soils affected by neutral sodium salts (mainly sodium chloride and sodium sulphate)

2. Soils affected by sodium salts capable of alkaline hydrolysis (mainly $NaHCO_3$, Na_2CO_3 and Na_2SiO_3).

In the course of the development of soil science and soil classification, two main groups of these soils have been distinguished: soils belonging to the first group have mainly been named saline, and those of the second group, alkali soils. These two main types differ not only in their chemical character but also in their geographical and geochemical distribution, as well as in their physical, chemical, physico-chemical and biological properties. The methods used for their reclamation and agricultural utilization are also different.

Although it is evident that in Nature the various sodium salts do not occur absolutely separately in soils, in most cases either the neutral sodium salts or the ones capable of alkaline hydrolysis exercise a dominating influence on soil forming processes and soil properties.

Accordingly, in the approach to defining the units of the World Map of Salt Affected Soils, two classes were recognized. These are:

A) A class dominated by chlorides and sulphates. This class is to be called: saline.

B) A class dominated by exchangeable sodium and/or by sodium bicarbonate and/or by sodium carborate. This class is to be called: alkali.
It is subdivided in:
a) a sub-class without structural B horizon
b) a sub-class with structural B horizon.

On the various continents, under the very wide range of environmental conditions, the general levels of the salinity or alkalinity of parent materials and groundwaters may sharply differ. The salinity or alkalinity tolerance of local crops varies widely, too. The potential salinity or alkalinity of an area depends, to a considerable extent, on the cropping system used in that particular area. It is more than obvious that in all these respects only very vague limit values may be given on a worldwide scale. Therefore it is necessary that - while keeping the basic principles in mind -, a certain flexibility be displayed in the definition of salinity and/or alkalinity limit values characterizing the salt affected soils of a given territory, that is, the local conditions should be also taken into consideration.

The very same principles were explained and applied with commendable clarity and consistency in the explanatory booklet for the map "Australian Soils with Saline and Sodic Properties".

It was recognized that in the case of Europe, where the dominating salt affected soils are the alkali soils, the division of class B into two sub-classes was not adequate to indicate important differences as they occur on this continent. Therefore the European Working Group agreed to subdivide class B into lower, more detailed grouping units, as follows:

a) A sub-class without structural B horizon
b) A sub-class with structural B horizon
    1. Solonchak-solonetz and calcareous solonetz
    2. Non-calcareous solonetz with an A horizon   <   15 cm
    3. Solodized and/or deeply leached solonetz and solod
    4. Solonized and slightly salt affected soils with minor structure formation.

While the further subdivision of class B was deemed absolutely necessary by the experts co-operating in the European Working Group, they agreed that the acid sulphate soils (thionic fluvisols according to the FAO soil classification of Europe, or known as "cat clays") should be omitted from the Map of European Salt Affected Soils. These soils occur in several coastal areas of this continent (for instance in Finland, Holland, Sweden, etc.) and differ fundamentally in their properties, genetics and environmental conditions from the common types of European salt affected soils.

Classes, Sub-classes and Mapping Units of European Salt Affected Soils.
Their Properties and Definitions

A) SALINE SOILS

On most continents the dominant types of salt affected soils are the saline soils. In Europe, however, the situation is different. On the basis of our present knowledge and of the available data it may be stated that the ratio between saline and alkali soils indicates the prevalence of the latter. The total extent of saline soils constitutes approximately less than a quarter of all the area covered by salt affected soils. Thus when the mapping of Europe is concerned, it is necessary to elaborate a more detailed grouping system for alkali soils, while a less detailed one is quite sufficient for saline soils. Accordingly, no subdivision of saline soils has been elaborated for the Map of European Salt Affected Soils.

Fig. 1. shows the schematic profile of a saline soil.

For saline soils the following definition was accepted: "Soils having a saline (salic) horizon within 125 cm below the surface (125 cm in case of coarse texture, 90 cm in case of medium texture and 75 cm in case of fine texture) or having an electric conductivity of more than 4 mmhos in at least some part of the soil within 25 cm below the surface; if the pH (H_2O 1:1) in this layer is 8.5 or less; an electric conductivity of more than 15 mmhos should occur within 125 cm in case of coarse texture, 90 cm in case of medium texture and 75 cm in case of fine texture below the surface."

In most cases the salt content of saline soils significantly surpasses the limit values given above.

In most European countries where salt affected soils occur, the percentile quantity of mineral residues determined from the 1 : 5 aqueous extract is used to establish the limit values of salinity. In the USSR and in several other countries this method is used almost exclusively.

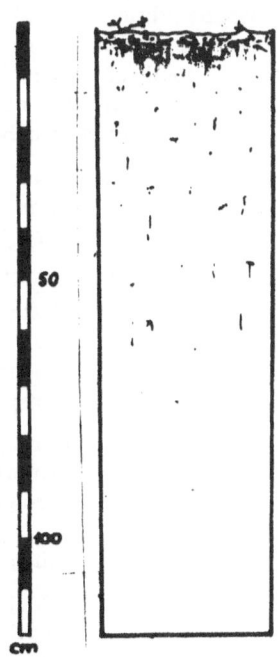

Fig. 1

Schematic profile
of a saline soil

It has been impossible to find exact correlation between the E.C. values and the data of the 1 : 5 aqueous extracts in relation to the diversity of the chemical composition of soluble substances in soils.

For rough estimation, however, .25% of salinity measured in the 1 : 5 aqueous extract may be considered as equivalent to an E.C. value of 4 mmhos/cm. Evidently, considerable deviations may occur in the case of different ions. If, in a given case, the determination of the exact correlation is necessary, methods for its calculation in case of different ions and concentrations may be found in: Jackson, M.L.: "Soil Chemical Analysis" (Englewood Cliffs. 1958.) or in: Darab, K. and Ferencz, K.:

13

"Öntözött területek talajtérképezése" (Soil Mapping and Control of Irrigated Areas) (OMMI. Budapest. 1969.)

The maximum of salt accumulation may be found at different depths in the soil profiles but very often it occurs in the top layer or near the surface. In Europe the saline soils have developed in the most arid regions. The few exceptions to this rule are caused by the salinity of local groundwaters or soil forming substrata.

Seasonal changes effected by climate and, particularly, by irrigation and drainage may occur in the salt content of saline soils, as well as in the distribution of salts in the different layers of the soil profile. These possible changes should always be taken into consideration when the soils are described and during the sampling and analytical procedures.

It is expedient to demonstrate the salt profiles of saline and alkaline soils.

Figures 2-6. represent the salt profiles of saline soils from different parts of Europe.

Figure 2. demonstrates the salt profile of a saline soil (chloride-sulphate-solonchak) from the Caspian Lowland, USSR.

Figure 3. represents the salt profile of a saline soil (solonchak) from the Romanian Danube Plain, Iancs Valley, west of Silistraru, Rumania.

Figure 4. demonstrates the salt profile of a saline soil (solonchak) from North Dobrogea, south of Lunca, Rumania.

The description of a saline soil (solonchak) profile, characteristic of the eastern part of the Romanian Danube Plain, is as follows:

A11sa   0- 3 cm: Very dark grey to very dark greyish-brown (10YR 3/1.5) silt loam, grey (10YR 5/1) when dry; weak, thin platy structure; very friable; slightly hard; plentiful roots; salt efflorescences; calcareous; clear wavy boundary.

A12sa   3- 17 cm: Texture and colours similar to the above horizon; weak, thin platy structure; very friable; slightly hard; very many efflorescences, veins and nests of soluble salts; grass roots; calcareous; gradual boundary.

ACsa   17- 27 cm: Dark grey to greyish brown (10YR 4/1.5), silt loam; grey to light brownish-grey (10YR 6/1.5) when dry; mottles of dark reddish--brown (5YR 3/4 moist) and yellowish-red (5YR 4/6 dry); massive; very friable; slightly hard; numerous thin veins of soluble salts; calcareous; abrupt boundary.

C1sa   27- 70 cm: Alluvium layers; very dark greyish-brown to dark greyish-brown (10YR 3.5/2) silt loam, grey (10YR 5/1) when dry; coarse textured layers have lighter colours; mottled with dark reddish--brown (5YR 3/4 moist) and yellowish-red (5YR 4/6 dry); massive; large amount of soluble salts; calcareous; abrupt boundary.

IIC2sa   70- 95 cm: Dark grey (10YR 4/1) silt loam, grey (10YR 5/0.5) when dry, mottles of reddish-brown (5YR 3/4 moist) and yellowish-red (5YR 4/6 dry); massive; salt crystals; calcareous; abrupt boundary.

IIIAB'sa 95-113 cm: Buried solonchak, black (10YR 2/1) silt loam, very dark grey to dark grey (10YR 3.5/1) when dry; few yellowish-red (5YR 4/6) mottles; massive; old roots; thin veins and nests of soluble salts; calcareous; gradual boundary.

14

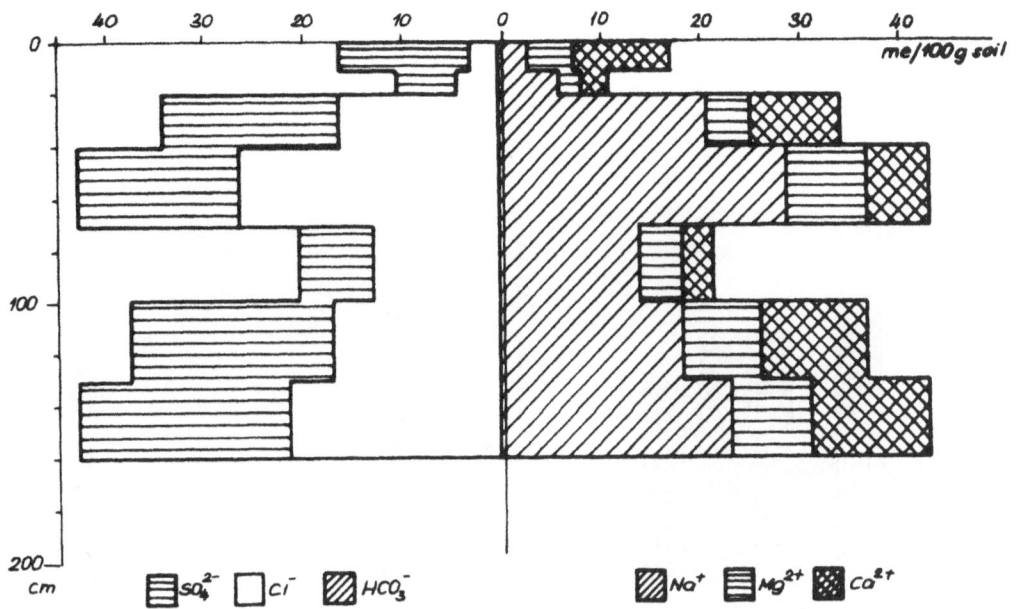

Fig. 2
Salt profile of a saline soil from the Caspian Lowland (chloride-
-sulphate-solonchak)

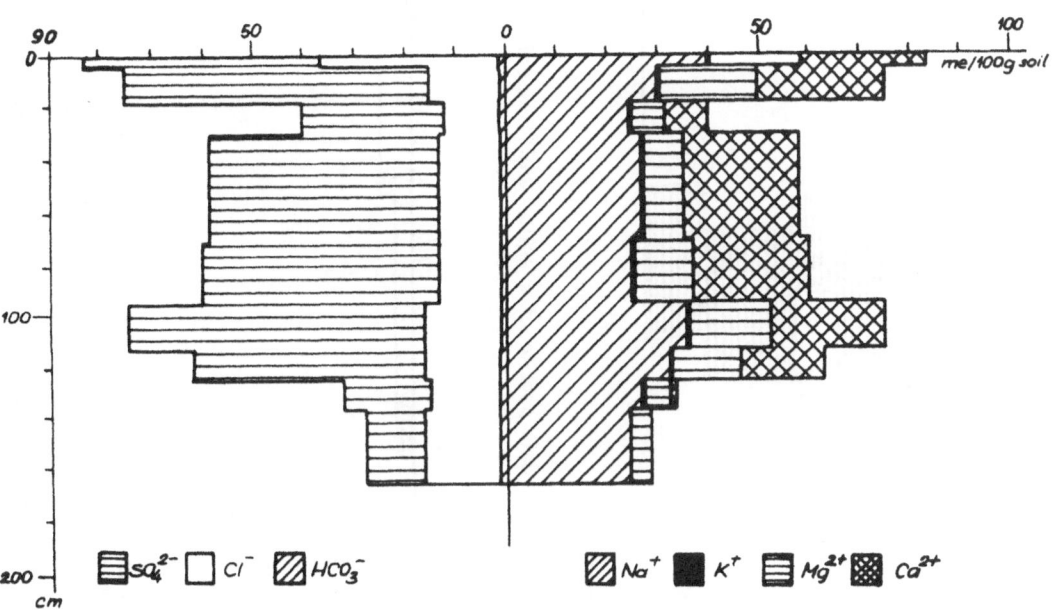

Fig. 3
Salt profile of a saline soil (solonchak) from the Romanian Danube
Plain, Iancs Valley, Rumania

15

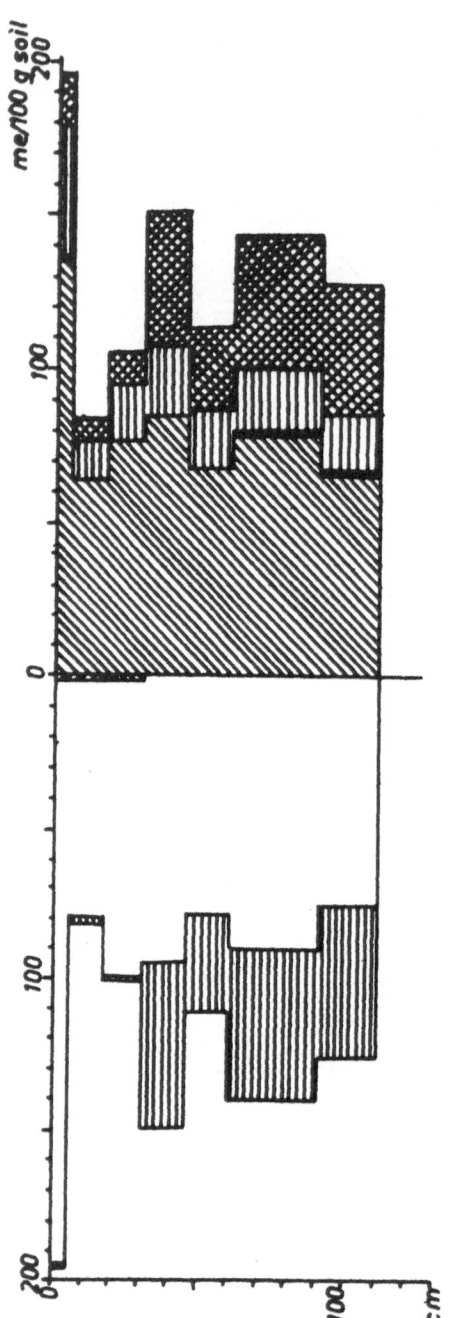
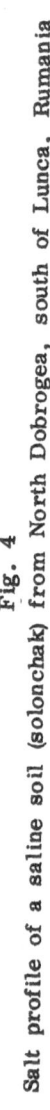

Fig. 4

Salt profile of a saline soil (solonchak) from North Dobrogea, south of Lunca, Rumania

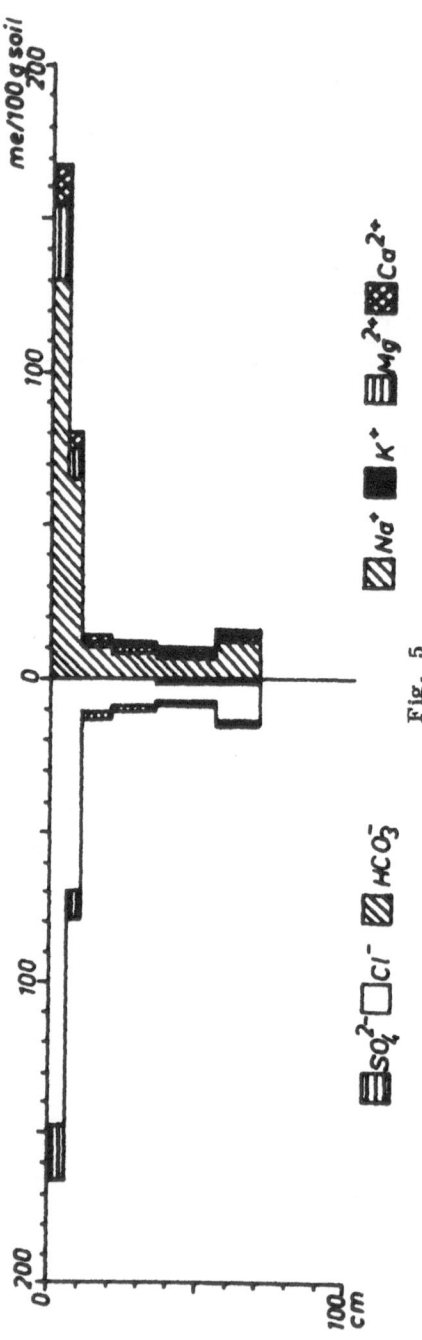

Fig. 5

Salt profile of a saline soil (marine solonchak) from the Black Sea shore, Chituc Peninsula, Rumania

16

IIIACb'sa 113-137 cm: very dark grey (10YR 3/1) silt loam grey (10YR 5/1) when dry (colours become lighter in the lower part of the horizon); yellowish-red mottles; massive; krotovinas; salt crystals; calcareous, abrupt boundary.

IVAb''gsa 137-165 cm: Another buried soil; dark grey (10YR 4/1) silt loam, grey (10YR 5/1) when dry; mottled; massive; thin roots; calcareous ; abrupt boundary.

IVCb''g 165-195 cm: Dark grey (10YR 4/1) silt loam, grey (10YR 5.5/1) when dry; dark reddish-brown (5YR 3/4) and grey (N5) mottles when moist and yellowish-red (5YR 4/6) mottles when dry; massive; calcareous.

Figure 5. demonstrates the salt profile of a saline soil (marine solonchak) from the Black Sea shore, Chituc Peninsula (Dobrogea), Rumania.

In general the profile of these soils is weakly developed, showing the following features :

A_{11}sa 2-8(5) cm: sand, loamy sand; black to dark greyish-brown (hue 10YR, with values of 2.5 to 3.5 and chroma less than 2) when moist and greyish-brown when dry (10YR 5-6/1); abundant fine iron spots; single grain; very friable, moist; slightly hard when dry; frequently small shell fragments occur; abundant thin roots; on surface white salt segregations; strongly calcareous; clear boundary.

A_{12}sa 6-24 cm thick; sand with some small shell fragments ; very dark greyish-brown, dark grey (hue 10YR with values of 3 to 4 and chroma less than 2) when moist and grey, greyish-brown (10YR 5/1--1.5) when dry; sometimes the colour may be olive-grey (5Y 3--4/2); abundant fine iron spots; other features similar to those above; clear boundary.

Ag 8-20 cm thick; sand; olive grey, pale olive (5Y 4.5-5.5/3) or brown (10YR 5/3) when moist and light grey (10YR-2.5Y 7/2) when dry; abundant grey and brown spots (mainly along the root channels); small shell frangments; other characteristics similar to those above; gradual boundary.

The G horizon occurs at 35-50 cm depth; loose sand; frequently very shelly; slightly cemented thin shell layers are to be found; it may be divided into two horizons:

G1: light brownish-grey (2.5Y 6/2) or grey, light grey to pale olive (5Y 6/1-3) when moist and light grey (2.5Y 7-7.5/2) when dry; abundant fine spots dark reddish-brown (5YR 3/4) or brown in colour; gradual boundary.

G2: mottled; moist; greenish-grey (5GY 6/1), olive grey or pale olive (5Y 5/2, 5Y 5/3) with abundant dark reddish-brown (5YR 2/2) and yellowish-red (5YR 4/6, 7.5YR 6/6) spots; sometimes when dry the colour becomes light grey, white (2.5Y 7.5/2, 10YR 8/1) with brownish-yellow (5YR 5/6) and pale brown (10YR 7/4) spots ; iron manganese segregations coating sand grains may be present; violent effervescence with diluted HCl.

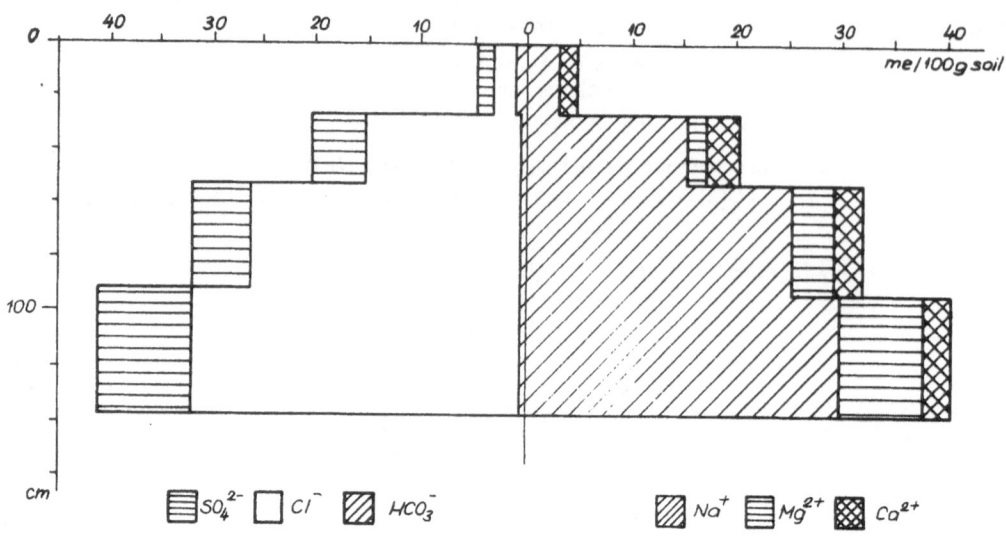

Fig. 6
Salt profile of a saline soil (sulphate-chloride-solonchak)
of Marismas, Spain

In Figure 6. the salt profile of a saline soil (sulphate-chloride-solonchak) of Marismas, Spain, may be seen.

The description of a saline soil of Marismas is as follows:

A_{11}	0- 5 cm:	Dark greyish-brown (10YR 4/2), calcareous fine clay, medium moderate laminar structure, abundant roots, slightly organic, moderately moist, very sticky and very plastic, abrupt and smooth boundary.
A_{12}	5- 25 cm:	Dark greyish-brown (10YR 4/2), calcareous fine clay, coarse moderate granular structure, abundant roots, very slightly organic, moist, very sticky and very plastic, gradual and smooth boundary.
AC	25- 95 cm:	Dark brown (10YR 4/3), calcareous fine clay, few roots, very coarse moderate prismatic structure, presence of slickensides, moist, very sticky and very plastic, common fine faint mottles about 5%, very dark grey (5YR 3/1) and dark reddish-grey (5YR 4/2), diffuse and smooth boundary.
Cy	95-135 cm:	Similar to the overlying layer but saturated with water; coarse yellowish-red patches, about 30% (5YR 5/8). Very few roots.
CG	135- cm:	Gley, greenish-grey (5GY 5/1). Water appears. No roots.

Elevation: 5 m above sea level.
Water table: 140 cm
Slope: 0%.

18

These profile descriptions show some morphological features of the different saline soils in Europe. As regards the morphology of saline soils in general, several common characteristics but also considerable differences may be observed.

The basic morphological precondition of saline soils is the lack of a structual B horizon. (See: Fig.1) Although several morphological systems use the letter "B", in those cases it never signifies a horizon distinguishable from the A horizon by its well-developed structural formation. Consequently, the profiles of saline soils are rather monotonous, from the surface down to the parent material. In a few cases, when saline soils have formed under bog-conditions, the top layers are humous, but usually, when the have developed under arid conditions, these soils are very poor in humus substances and their humus content is lower than 1%. The low plant nutrient content (mainly N and P_2O_5) is also characteristic of most saline soils.

The high salinity determines practically all physical and chemical properties of saline soils, consequently, when these properties are evaluated, primarily the salt content of the soils and its influence should be taken into consideration.

Many classification systems employ the term "chloride and/or sulphate solonchak" for saline soils. (e.g. see Table 1. on p. 39.)

B) ALKALI SOILS

In alkali soils the presence of Na-salts capable of alkaline hydrolysis det rmines the soil properties. Due to their effect either the high alkalinity of the soil solution hinders plant growth, or the alkalinity renders the physical soil properties disadvantageous for the water supply of plants. Evidently, often both of these processes exert their harmful influence, though in alkali soils without structural B horizon the former, and in alkali soils with structural B horizon the latter dominates.

As a rule, in the case of soils belonging to sub-class a) (alkali soils without structural B horizon) a considerably high concentration of water soluble sodium salts capable of alkaline hydrolysis can be found even in the top layers. In sub-class b) (alkali soils with structural B horizon) the concentration of water soluble salts is often very low and, except in the B horizon, alkalinity may also be quite moderate. Owing to the considerable differences between the properties of soils belonging to sub-classes a) and b), their characteristics and definitions should be discussed separately.

a) Alkali soils without structural B horizon

In these soils a saline (salic) horizon can be found within 125 cm (125 cm in case of coarse texture, 90 cm in case of medium texture and 75 cm in case of fine texture) below the surface, or the electric conductivity is more than 4 mmhos in at least some part of the soil within 25 cm below the surface. The pH (H_2O 1:1) should be more than 8.5 somewhere in the 0-25 cm thick layer. (It should be noted that if the pH is determined against phenolphthalein, a careful analytical procedure is necessary because the finely dispersed calcium carbonate also gives a slight pink colour with phenolphthalein. In such cases a more precise pH determination or titrimetrical analysis must be carried out.)

As a rule, in alkali soils without structural B horizon a considerably high concentration of sodium salts capable of alkaline hydrolysis - mainly of sodium carbonate -, may be found. Of all these salts which commonly occur in soils, sodium carbonate exercises the most harmful effect on both soils and plants. Thus this sub--class represents salt affected soils which have very disadvantageous properties for agriculture. Their fertility, if any, is very low. Not only the alkalinity but, in most cases, also the salinity of these soils is rather high. This is why in many classification systems they are denominated "alkali-saline" or "saline-alkali" soils. Although sometimes neutral sodium salts may also prevail among the water soluble substances in these soils, nevertheless the dominant role is played by sodium carbonate owing to its high alkalinity.

While in soils belonging to class A (saline soils) the high salt concentration occurs mainly under arid conditions and only in a few cases may it be found in other climatic zones, soil salinity caused by sodium carbonate occurs in arid, moderate, or even in humid climate as well. This phenomenon can be explained by the differences in the geochemical and biogeochemical processes involved in the accumulation in soils of neutral sodium salts and sodium salts capable of alkaline hydrolysis. Accordingly, the occurrence of saline soils is characteristic mainly of deserts, semi-deserts and very dry steppe regions, while alkali soils without structural B horizon may be found in various climatic zones.

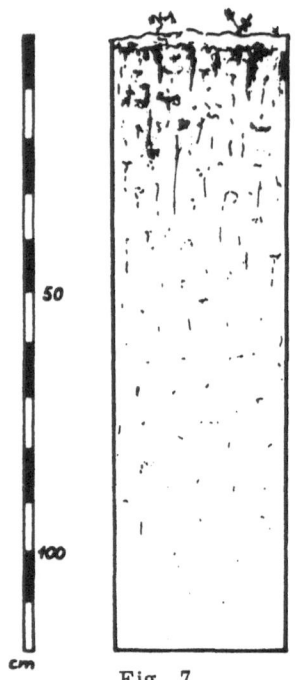

Fig. 7
Schematic profile of
an alkali soil without
structural B horizon

Figure 7. demonstrates the schematic profile of an alkali soil without structural B horizon.

Similarly to the profile of a saline soil (Fig. 1) this soil also lacks a B horizon of well developed structural elements, distinguishable from the A horizon. On the basis of this morphological similarity, in many classification systems these soils are indicated as solonchak soils. In order to distinguish them from saline soils, they are named sodium carbonate or soda solonchaks as distinct from sodium chloride or sodium sulphate solonchaks.

The percentage of sodium carbonate and bicarbonate may vary from a few tenth up to several percent in these soils. Depending on local conditions, the salt maximum may occur at the surface or in the deeper layers. The depth of the water table is different but in Europe, in most cases, the groundwater is within 2 meters of the surface.

As a rule, the more we approach the moderate or humid climatic zone, the nearer rises the groundwater to the surface.

It has been found that in many cases there exists a close correlation between the chemical composition of the salt content of the soil profile and that of the groundwater, clearly proving the influence of mineralized groundwaters on soil formation.

Due to the high alkalinity of these soils, the top layers are rather compact, structureless, and their extremely low water permeability, or virtual impermeability is a very important factor from the point of view of drainage or chemical reclamation.

Most alkali soils without structural B horizon are very poor in humus and plant nutrients (particularly in N and P). Humous top layers may occur, however, more often in humid or semi-humid areas than in the case of desert soils, owing mainly to bog conditions or comparatively wet conditions. Especially when the soda forming processes are associated with temporary waterlogged conditions, the humus content of the top layer may reach several percent, while in other cases it amounts only to a few tenth of one percent.

20

Because high alkalinity is accompanied by high salinity, the former governs not only the chemical but also the physico-chemical processes in these soils. Consequently, the determination of exchangeable cations is not always necessary and quite often it may even prove analytically difficult during the survey and routine analysis of alkali soils without structural B horizon.

If the concentration of sodium salts capable of alkaline hydrolysis decreases, transitional types between alkali soils with and without structural B horizon may develop. This transitional form is known in several classification systems as "solonchak-solonetz" or "solonetz-solonchak". Similar process may develop during the desalinization of saline soils.

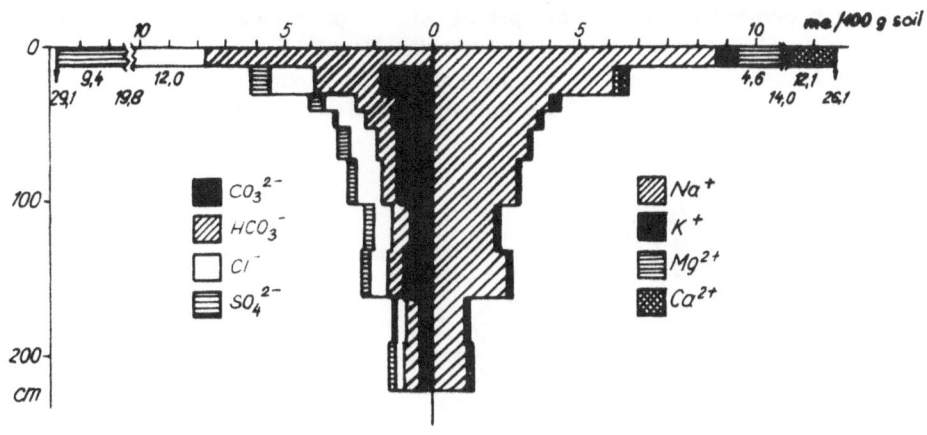

Fig. 8
Salt profile of an alkali soil without structural B horizon from the
Hungarian Danube Valley, Apaj-4., Hungary

In the following the description of an alkali soil without structural B horizon is given. (From the Danube Valley in the Hungarian Lowland. Profile - Apaj-4.)

Surroundings: Pasture of extremely poor quality with bare spots.
Topography: Flat.
Vegetation: Lepidium cartilagineum, Camphorosma ovata, Artemisia monogyna, Atropis (Puccinellia) limosa, Nostoc commune.
Depth of the profile: 120 cm.
Effervescence (with dilute acid): To the surface.
Alkalinity against phenolphthalein: To the surface.
Thickness of the humus layer: 12 cm.
Water table: 64 cm.
Genetic horizons:

A   0- 12 cm: Gray (pale, mouse-gray, whitish when dry), slightly moist, sandy silt. Undeveloped, weak structure. Relatively plentiful roots. Gradual boundary.

B_1 12- 30 cm: Light gray, slightly moist, silty loam. Slightly prismatic structure (especially when dry). Gradual boundary.

B_2 30- 52 cm: Whitish gray, loamy silt. Moist. Prismatic structure when dry. Horizon of lime accumulation. Abrupt boundary.

21

C   52-120 cm: Gray, slightly silty sand. Loose. There are few, small gravels.
                 Large white silty spots and gray, very loose sandy spots. The
                 silty spots get rarer with depth.

<u>Soil type:</u>   Sodic solonchak on calcareous Danube alluvial sand. (Alkali soil
                 without structural B horizon.)

(<u>Remark</u>: as it was mentioned earlier, in the Hungarian system of profile description symbol "B" is used to indicate the horizon below the A horizon even if it is
structureless and cannot be clearly distinguished by a sharp boundary from the A
horizon.)

Figure 8. demonstrates the salt profile of the previously described alkali soil
without structural B horizon from the Hungarian Danube Valley, Apaj-4.

Figure 9. shows the chemical composition of the groundwater, corresponding ob-
viously to the salt composition of the soil described above, Apaj-4.

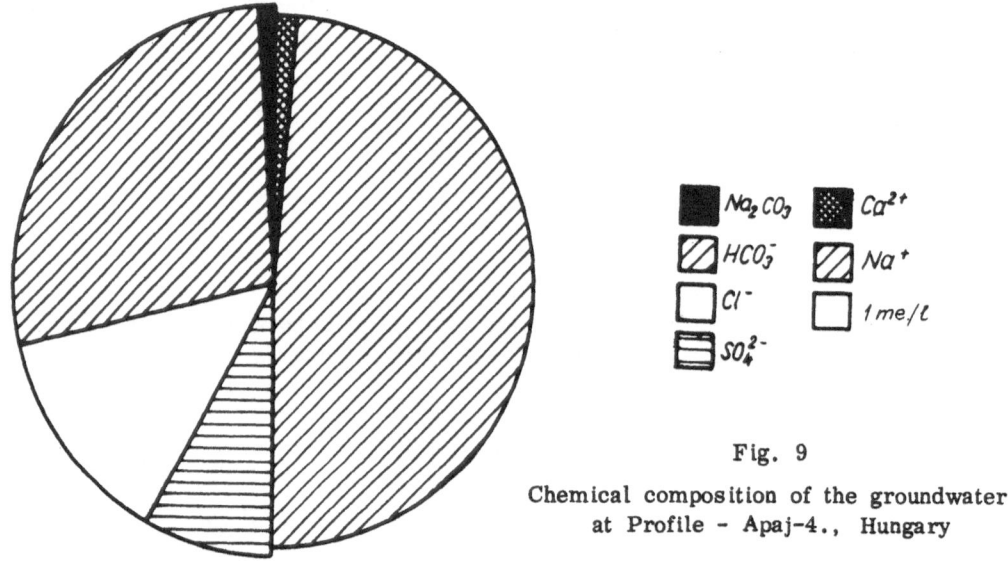

Fig. 9

Chemical composition of the groundwater
at Profile - Apaj-4., Hungary

b) Alkali soils with structural B horizon

The definition of these soils is based on the prismatic or columnar structure of
the B horizon which is accompanied by a high percentage of exchangeable sodium
ions (ESP > 15).

This sub-class represents the most widespread group of salt affected soils in
Europe.

In some classification systems soils belonging to this sub-class are denominated
"solonetz" and "solonetz-like" ("solonetzic", "solodized", "solod", etc.) soils. (See
the Correlation Table on page 40.). There are countries where various, sometimes
absolutely different names are used to indicate them, and also in the scientific litera-
ture the term "solonetz" is interpreted in many different ways, therefore it is necessary to summarize briefly our knowledge of the formation, morphology and properties of these soils.

In Figure 10. schematic profiles of alkali soils with structural B horizon (solonetz) are shown.

These soils always have a structural B horizon in their profiles, which, as a rule, has a well-developed structure, mainly columnar. It can be easily distinguished from the horizon (A horizon) above it, which is less compact and whose structure is less developed. This B horizon determines the genetic type of these soils, their main

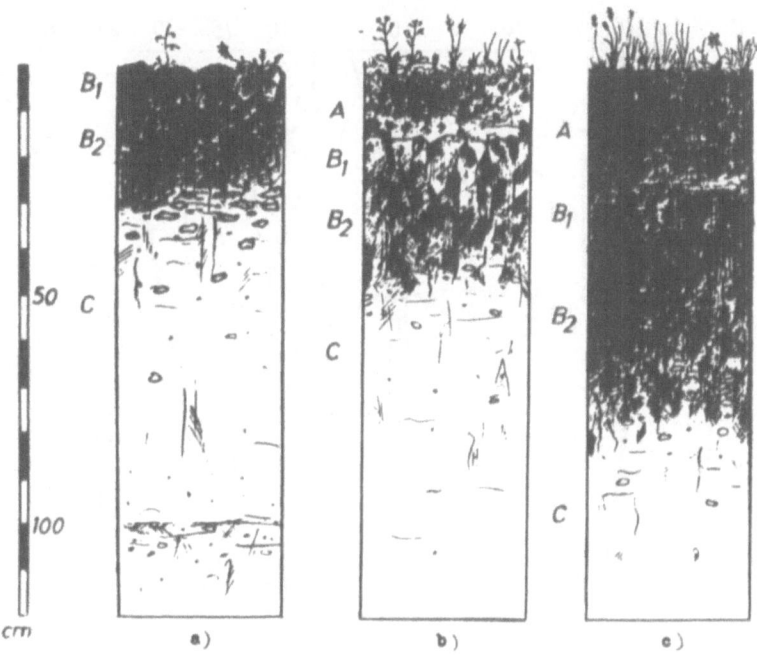

Fig. 10

Schematic profiles of alkali soils with structural B horizon
(solonetz)

physical, chemical, physico-chemical and biological properties, as well as their fertility together with the possibilities of their agricultural utilization.

The structural B horizon is situated at various depths, depending on local circumstances. In some cases it is at the surface (the A horizon completely lacking). (See: Figure 10.a)

The structural B horizon always markedly differs from the A horizon not only in morphology, colour and structure, but also in its physical, chemical, physico-chemical and biological properties. Figure 11. schematically represents some of the chemical, physical and physico-chemical properties of an alkali soil with structural B horizon. It demonstrates that at a given depth below the surface (in this case at about 30 cm) an illuvial horizon, that is an accumulation horizon, may be found. This is named B or B_1 horizon. In this horizon the accumulation of clay particles and sesquioxides may be observed and the water soluble organic matter content as well as the ESP values show their maximum, while the ratio of $SiO_2 : R_2O_3$ is the lowest, silicon

23

compounds are comparatively at a minimum. In Figure 11. this horizon is between 20-30 cm, as is often the case in nature, but it frequently occurs at depths other than this. Naturally in those cases the respective maximum and minimum values may be found at the depth where the B horizon developed.

It may be seen in Figure 11. that - as regards the movement and accumulation of materials - in the A horizon the situation is just the contrary. The A is the eluvial horizon

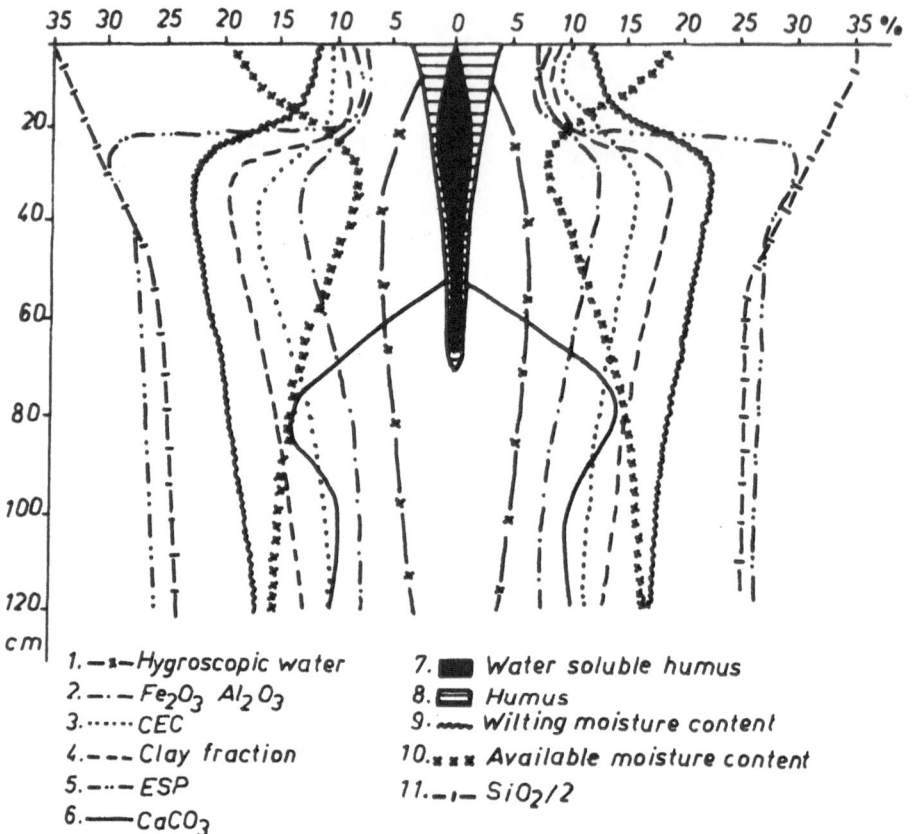

1. —ı—Hygroscopic water
2. —·—Fe$_2$O$_3$ Al$_2$O$_3$
3. ······ CEC
4. ----Clay fraction
5. —··—ESP
6. ——CaCO$_3$

7. ▆ Water soluble humus
8. ⬭ Humus
9. ∿ Wilting moisture content
10. ×××Available moisture content
11. —ı—SiO$_2$/2

Fig. 11

Schematic representation of some of the physical, chemical and physico-chemical properties of an alkali soil with structural B horizon (solonetz). (After Kovda, in modified form)

The thickness of the A horizon is a very important feature of these soils. Their fertility is in direct proportion to the thickness of their A horizons. For one thing, it determines the amount of water retained and available for plants. The B horizon is much more compact than the A and the penetration of plant roots into it is always impeded. Besides - as clearly shown in Figure 11 - the amount of exchangeable sodium and water soluble sodium salts harmful to plants is lower in the A than in the B horizon.

It is obvious that the thickness of the A horizon has always been taken into consideration in the grouping and classification of solonetz soils. Naturally the limit values of the thickness of the A horizon - on the basis of which the solonetz soil is included in a certain group - depend to a large extent on local conditions. Depending on climate, geochemistry, plant nutrition and other factors, sometimes thicker and sometimes thinner A horizons correspond to a given water and nutrient economy of the soil profile. The following grouping is intended only to give a general directive with regard to the division of solonetz soils into three groups according to the thickness of the A horizon.

Name	Thickness of the A horizon
Shallow solonetz soil	0 - 6 cm
Medium solonetz soil	7 - 16 cm
Deep solonetz soil	> 16 cm

In European countries where alkali soils with structural B horizon are widespread, this, or similar grouping systems are adopted for both survey and reclamation purposes.

One of the important characteristics of the B horizons of solonetz soils is their high exchangeable sodium content. It is generally held that the high exchangeable sodium content is responsible for the poor physical and water regime properties and the compact structure of the B horizons of these soils. The exchangeable sodium content is usually expressed in me/100 g soil or, even more frequently, in percentage of the cation exchange capacity (ESP). As it was mentioned earlier, the limit value of ESP for alkali soils with structural B horizon is 15; when the ESP value is about 5 - 7, the first signs of the development of the compact B horizon may be observed in the profile. Naturally, the limit values are approximate and may slightly vary depending on soil properties and local conditions. The basic feature of identification for these soils is always the above described morphology of the B horizon.

Depending on local circumstances, some alkali soils with structural B horizon may have considerable amounts of water soluble sodium salts even in the upper layers (more than 4 mmhos) while others are practically devoid of salt in the entire profile. Of the water soluble sodium salts sometimes bicarbonates, sometimes sulphates or even chlorides prevail. With regard to the maximum accumulation of water soluble salts in the profile, as a rule, it occurs in the lower part of B_2 horizons but, depending on the conditions of soil formation, it may also occur in other layers, either above or below the B horizon. The pH value of these soils may also vary to a considerable degree. In some cases a strongly alkaline pH may be observed from the surface, sometimes the pH of the top layer is neutral or even slightly acid, and there are solonetz soils in which a strongly alkaline pH does not occur at all anywhere in the profile. In the B horizon, however, where the maximum exchangeable sodium percentage (ESP) may be found, the pH is always over 7.

In the B horizon the sodium ions are mainly in exchangeable form, adsorbed on the soil colloids the maximum of which also occurs there. These exchangeable sodium ions - depending on the dynamics of the equilibrium conditions between the solid and liquid phases - are capable of alkaline hydrolysis. This phenomenon is influenced by the CEC and ESP values, by the chemical composition and concentration of the soil solution, especially by the CO_2 tension, and by many other factors. Because of this, the alkaline hydrolysis of the soil colloids, saturated with sodium to a higher or lower degree, results in more or less alkaline conditions in this horizon. It is evident that - with the exception of solonetz soils affected by strongly developed solod forming processes - in the B horizon, where sodium compounds capable of alkaline hydrolysis play a dominant role, alkaline hydrolysis usually takes place, due to the interaction of the solid and liquid phases of the soil.

As regards the formation of alkali soils with structural B horizon, great importance should be attached to the influence of the environmental conditions - particularly to that of mineralized groundwaters - on soil forming processes. It was on the basis of this principle that sub-class b) was further divided into four grouping units. (See on page 12.)

1. Solonchak-solonetz and calcareous solonetz, and
2. most of the non-calcareous solonetz soils with an A horizon < 15 cm in thickness

- have been forming in permanent link with mineralized groundwaters, while

3. solodized and/or deeply leached solonetz and solod, and
4. solonized and slightly salt affected soils with minor structure formation
- have developed either without being linked with mineralized groundwaters, or - if linked -, the groundwaters being practically non-mineralized.

In the following the characteristics and the definitions of the four groups of sub-class b) of alkali soils are outlined.

1. Solonchak-solonetz and calcareous solonetz soils.

Solonchak-solonetz soils have an electric conductivity of more than 4 mmhos in at least some part of the profile within 25 cm below the surface as well as a slightly developed structural B horizon also within 25 cm of the surface. Calcareous solonetz soils have well-developed structural B horizons at a certain depth and are calcareous within 25 cm below the surface. They show the characteristic morphological features and have the corresponding limit value of ESP.

Below these soils, just like in the case of soda-solonchak soils, the mineralized groundwater may be found within 1 meter of the surface in most parts of Europe. The pH (H_2O 1:1) should be more than 8.5 somewhere in the 0 - 25 cm layer, also like in soda-solonchak soils. In Europe these soils occur in association with alkali soils without structural B horizon and they have many features (e.g. humus content, nutrient content, physical and water regime properties) in common.

As regards the salt content of calcareous solonetz soils, their total salinity is less than in solonchak-solonetz soils and the salt maximum appears below the surface layer.

Within sub-class b) calcareous solonetz soils are attached to solonchak-solonetz soils on the basis of the numerous similarities in their properties, as well as on that of the many transitional formations between the two types.

In the following the description of a calcareous meadow solonetz soil from Apaj in the Danube Valley of the Hungarian Lowland is given. (Profile - Apaj-3.)

Surroundings: Pasture of poor quality, spotted with salt affected soil.
Topography: Slightly uneven, flat.
Vegetation: Festuca pseudovina, Cynodon dactylon, Ononis spinoza, Centaurea sp., Potentilla sp. The depressions are covered with rich grass and species characteristic of swampy areas. These slight depressions are surrounded by salt affected soils. On these soils salt efflorescences and encrustations may be found. Their vegetation consists mainly of Camphorosma ovata, Festuce pseudovina and, at several places, Nostoc commune.
Depth of the profile: 120 cm.

<u>Effervescence (with dilute acid)</u>: To the surface.
<u>Alkalinity against phenolphthalein</u>: at several places to the surface.
<u>Thickness of the humus layer</u>: 30 cm.
<u>Water table</u>:   210 cm.
<u>Genetic horizons</u>:

A   0- 2 cm: Pale gray, slightly moist, fine granular structure, many fine roots, sod. Abrupt boundary.

B₁ 2- 11 cm: Gray, dry, very hard, columnar structure, sandy loam. Relatively plentiful roots. Abrupt boundary.

B₂ 11- 26 cm: Brownish gray, slightly moist, prismatic structure, loam. Plentiful roots. Slightly hard. Light gray mottles at several places. Gradual boundary.

BC 26- 43 cm: Whitish gray, slightly moist, hard, platy, prismatic structure, silty loam. Vertical dark humus tongues and gray silt mottles. Horizon of lime accumulation. Abrupt boundary.

C₁ 43- 95 cm: Yellow, slightly moist, silty sand. Vertical humus streaks along the roots. Remains of snails. The horizon gets gradually sandy with depth. Slightly rusty mottles of iron from 75 cm. Abrupt boundary.

C₂ 95-200 cm: Yellowish gray, slightly moist, silty sand. Lots of rusty iron mottles. The horizon gets gradually more sandy, moist and loose.

    200-   cm: Gray, very loose, wet sand.

<u>Soil type</u>:   Salinized calcareous shallow meadow solonetz on calcareous Danube alluvial sand. (Alkali soil with structural B horizon.)

Figure 12. demonstrates the salt profile of the Apaj-3 soil profile.

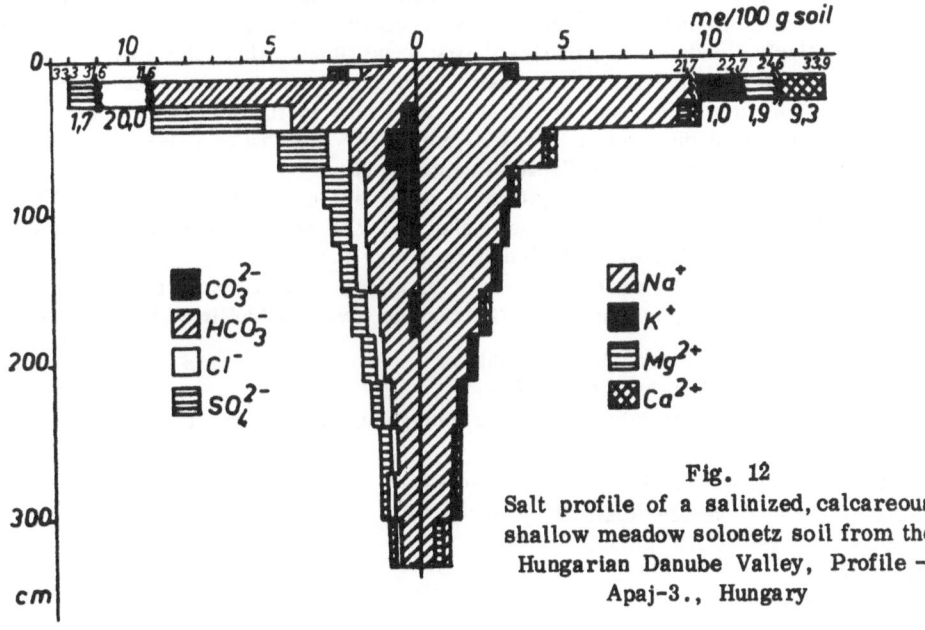

Fig. 12
Salt profile of a salinized, calcareous shallow meadow solonetz soil from the Hungarian Danube Valley, Profile – Apaj-3., Hungary

As mentioned before, the mineralized groundwater exerts a decisive influence on the formation of these soils, as it can be clearly seen in Figure 13. showing the chemical composition of the groundwater below the Apaj-3 profile.

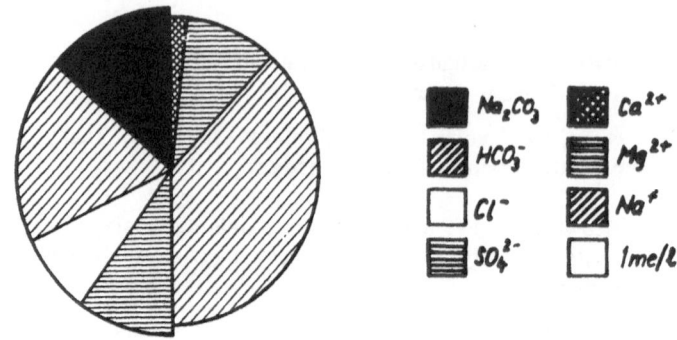

Fig. 13
Chemical composition of the groundwater at Profile - Apaj-3.,
Hungary

2. Non-calcareous solonetz soils with A horizons less than 15 cm in thickness.

These soils have a well-developed B horizon in their profiles below the A horizon which is less than 15 cm in thickness. In the case of the so-called crusty solonetz soils the A horizon is completely lacking and the B horizon is on the surface. The soil is non-calcareous in the 0 - 25 cm layer. Sometimes the profile is free of carbonate within 1 meter, or even more, below the surface but in most cases calcium carbonate appears at about 0.5 meter. Parallel with the appearance of the structural B horizon the ESP value shows its maximum - always more than 15 - in that horizon.

Depending on the local environmental conditions the water table may be permanently or temporarily linked with the soil profile; it may also happen, however, that - if the water table is situated at a greater depth - they are never in connection.

Accordingly, under comparatively dry conditions, when the soil profile is not linked with the mineralized groundwater, so-called steppe solonetz soils develop. This formation is widespread in Europe, it occurs in the steppe-regions, where the annual precipitation does not reach 400 mm (e.g. the Trans-Volga territory and the dry regions of Ukraine in the USSR, Rumania, etc.)

While most part of the so-called meadow solonetz soils which are linked with the groundwater belongs to this group, the majority of soils not, or only temporarily linked with the groundwater is included in group 3.

In the following the description of a meadow solonetz soil from the Hortobágy region of the Tisza River Valley in Hungary is given. (Profile - Hortobágy-2.)

Surroundings: Pasture of poor quality.
Topography: Slightly uneven, flat.
Vegetation: Artemisia monogyna, Polygonum aviculare, Festuca pseudovina.
Depth of the profile: 120 cm.
Effervescence (with dilute acid): from 32 cm.
Alkalinity against phenolphthalein: from 50 cm.
Thickness of the humus layer: 56 cm.
Water table: 230 cm.
Genetic horizons:

A	0- 3 cm:	Pale gray, slightly moist, loose, ash-like feel, sandy loam with many fine roots. Abrupt boundary.
B_1	3- 15 cm:	Gray, dry, extremely hard, distinctly columnar structure, clayey loam. Relatively plentiful roots. The tops of columns and, at some places, the sides are discoloured, solodized. Abrupt boundary.
B_2	15- 31 cm:	Gray, somewhat darker, slightly moist and hard, fine prismatic structure, clayey loam. Moderate amount of roots. In calcium carbonate abrupt, otherwise gradual boundary.
B_3	31- 55 cm:	Brownish gray, somewhat lighter in colour, slightly moist and hard, fine prismatic structure, clayey loam. Few roots. Iron mottles and iron concretions growing more frequent with depth. White lime spots, fine lime concretions. Abrupt boundary in colour.
C_1	55- 93 cm:	Grayish yellow, slightly moist and moderately hard, coarse prismatic structure, loess-like clayey loam. Dark, clayey humus streaks, white lime mottles, lots of lime concretions, stains of iron, soft iron concretions. Gradual boundary.
C_2	93-110 cm:	Grayish yellow, slightly moist, moderately hard, loess-like clayey loam. Lime and iron concretions, rusty stains of iron. Greenish gray mottles of gley.
	110-130 cm:	Yellow loamy clay, lots of lime and iron concretions.
	130-150 cm:	Yellow silty clay, lime and iron concretions.
	150-180 cm:	Yellowish gray clay, segregated lime and lots of iron concretions.
	180-210 cm:	Gray clay. Some segregated lime and iron.
	210-220 cm:	Reddish brown, sticky clay.

Soil type: Shallow meadow solonetz on calcareous, loess-like clayey loam. (Alkali soil with structural B horizon.)

Figure 14.demonstrates the salt profile of Hortobágy-2.

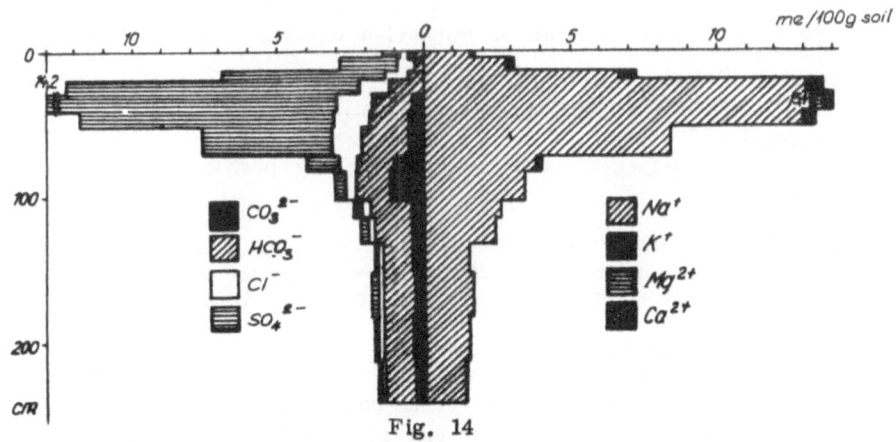

Fig. 14

Salt profile of a shallow meadow solonetz soil from the Hortobágy region, Profile - Hortobágy-2., Hungary

In order to illustrate the influence of the mineralized groundwater on the composition of the water soluble salts in the Hortobágy-2 profile, the chemical composition of the groundwater is shown in Figure 15. A close correspondence may be observed between the composition of the water soluble salts and that of the mineralized groundwater.

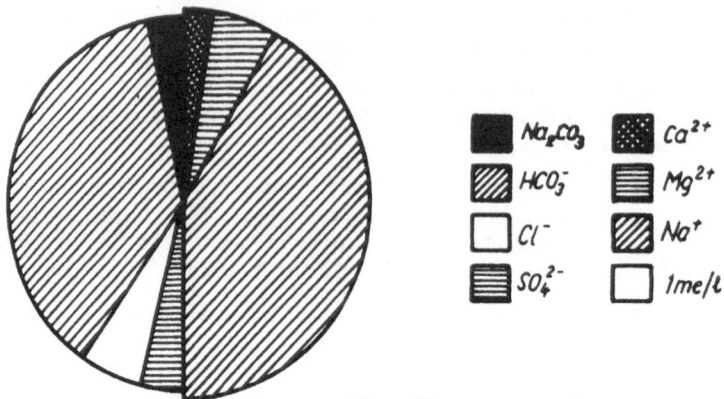

Fig. 15

Chemical composition of the groundwater at Profile - Hortobágy-2., Hungary

3. Solodized and/or deeply leached solonetz and solod soils.

A deeply leached solonetz soil has, as a rule, an A horizon thicker than 15 cm. These soils are never calcareous in the upper 0 - 25 cm layer and are frequently non-calcareous down to a depth of 60 cm or more. As regards the salt profile of these solonetz soils, the total amount of water soluble salts is less in the upper horizons and the maximum of salt accumulation occurs considerably deeper in the profile than in the case of solonetz soils belonging to group 2.

When a temporary link exists between the solonetz profile and the groundwater, the so-called solonetz soil turning into steppe formation develops.

In Figures 16 - 17 the salt profiles of two solonetz soils turning into steppe formation in the USSR are shown. (After Dimo and Yulidov)

In the following the description of a solonetz soil turning into steppe formation is given. (From the eastern part of the Hungarian Lowland, Profile - Hosszuhát-5.)

Surroundings: Pasture of poor quality.
Vegetation: Festuca pseudovina, Achilles millefolium.
Depth of the profile: 132 cm.
Effervescence (with dilute acid): none at any part of the profile.
Genetic horizons:

A   0- 19 cm: Pale gray, weak, fine granular structure, dry loam. Plentiful roots in the upper 3 cm. Moderate roots in the whole layer. Fungoid growth at some places along the roots. Abrupt boundary.

B₁ 20- 67 cm: Dark gray, almost black, very hard, compact columnar structure, clay. Vertical fissures in the whole layer. Root remnants occur more often down to 37 cm, below that there are only a few. At the bottom of the layer lots of reddish-yellow iron mottles. With depth the columnar structure gradually turns into columnar-prismatic structure. Gradual boundary.

B$_2$ 68-102 cm: Lighter gray than the former layer, prismatic structure, very compact clay. Slightly moist. With depth the colour takes on a yellowish hue, the prismatic structure becomes weaker and gradually moister. Below 76 - 77 cm plentiful bluish gray gley - and reddish yellow iron mottles, and a few small soft iron concretions. At the bottom of the layer, along old roots gypsum veins at some places. Gradual boundary.

C 103- cm: Yellowish clay, plentiful bluish gray gley- and reddish yellow iron mottles. Bluish black Mn mottles. Moister than the overlying layer, with depth gradually becomes moister.

<u>Soil type:</u>   Columnar deep meadow solonetz on clay. (Alkali soil with structural B horizon.)

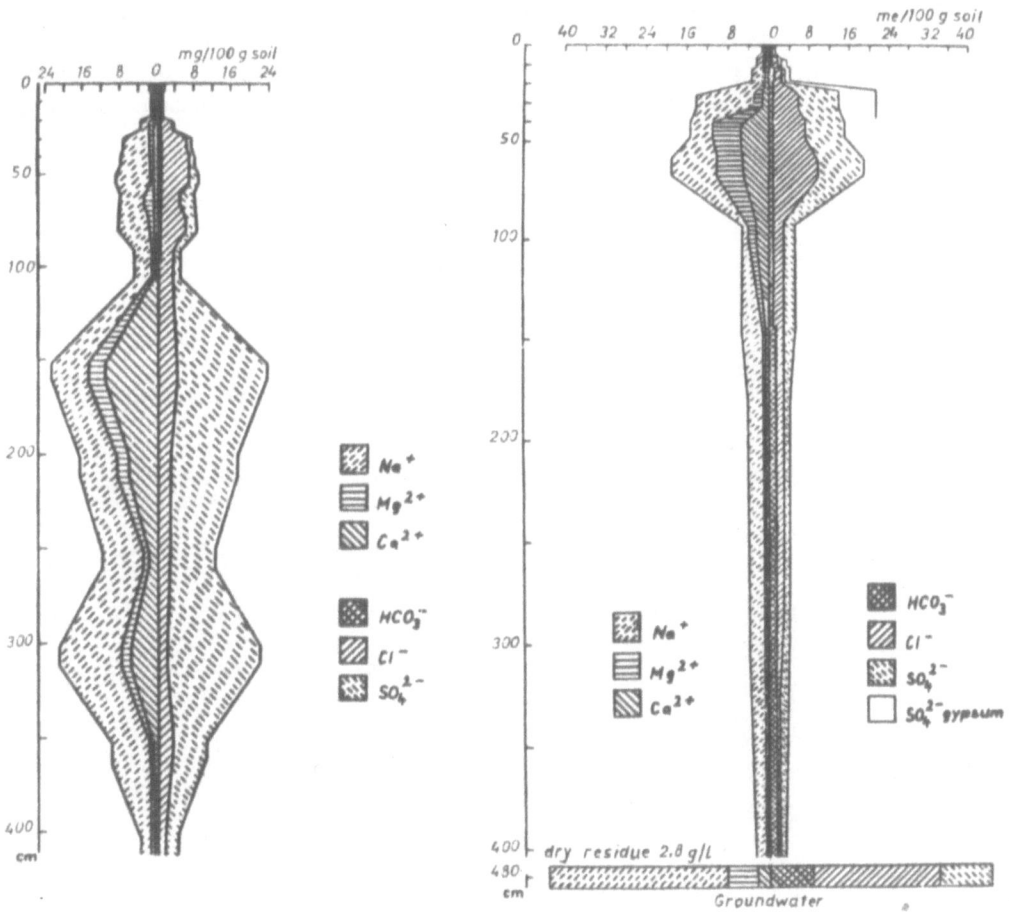

Fig. 16
Salt profile of a solonetz soil
turning into steppe formation
in the USSR. (After Dimo)

Fig. 17
Salt profile of a solonetz soil
turning into steppe formation
in the USSR. (After Yulidov)

Figure 18. demonstrates the salt profile of Hosszuhát-5.

Solonetz soils turning into steppe formation are widespread in practically all European countries where solonetz soils occur.

In the case of steppe solonetz soils the soil profile is practically never linked with the groundwater.

In Figure 19. the salt profile of a steppe solonetz soil from the USSR is presented (After Kovda). In the lower part of this Figure the chemical composition of the groundwater is also demonstrated.

The main part of steppe solonetz soils belongs to this group but certain steppe solonetz soils having an A horizon less than 15 cm in thickness are included in group 2.

The signs of solodization may be found in many of these soils, in the A horizons. These signs are as follows: an expressed pale gray colour and lamellated and powdery structure of the A horizon.

Solodization is, in most cases, closely associated with the formation of the A horizon in solonetz soils.

In solod soils the formation of the A horizon is very expressed, and the amount of the so-called KOH soluble SiO_2 exceeds 1 percent in the A horizon. In Europe

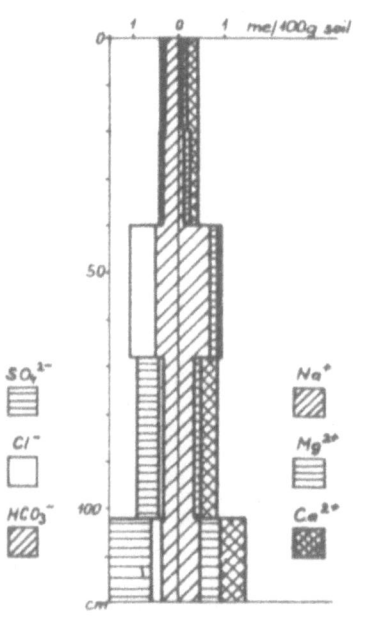
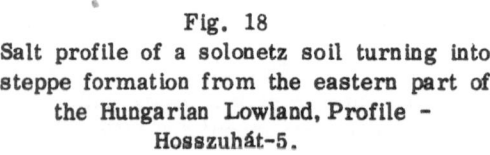

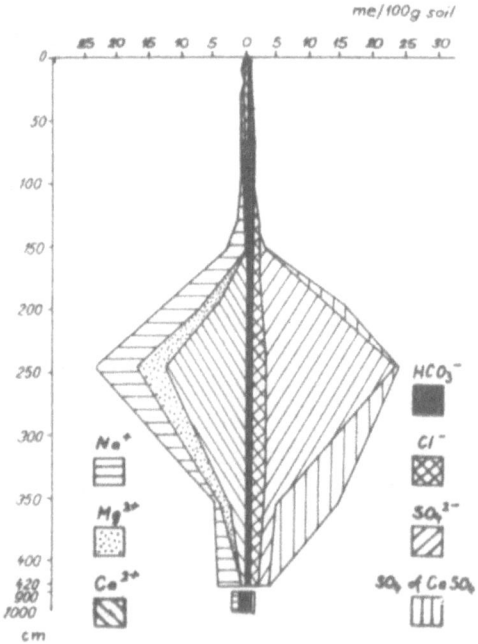

Fig. 18

Salt profile of a solonetz soil turning into steppe formation from the eastern part of the Hungarian Lowland, Profile – Hosszuhát-5.

Fig. 19

Salt profile of a steppe solonetz soil from the USSR. (After Kovda) solod soils are closely associated with solodized solonetz soils. In a very few cases solod soils occupy areas of considerable extent but usually they occur in small spots. This is the reason why they cannot be represented separately on the Map of European Salt Affected Soils prepared on the scale 1 : 5,000,000.

The schematic profiles of solod soils may be seen in Figure 20.

With regard to the morphology of a solod profile, two cases should be distinguished. Figure 20.a) represents a schematic solod profile, when the accumulation of SiO_2 is on the surface. In this case the A horizon is the solodized one. This occurs, as a rule, if the groundwater is capillarily linked with the upper layers of the soil

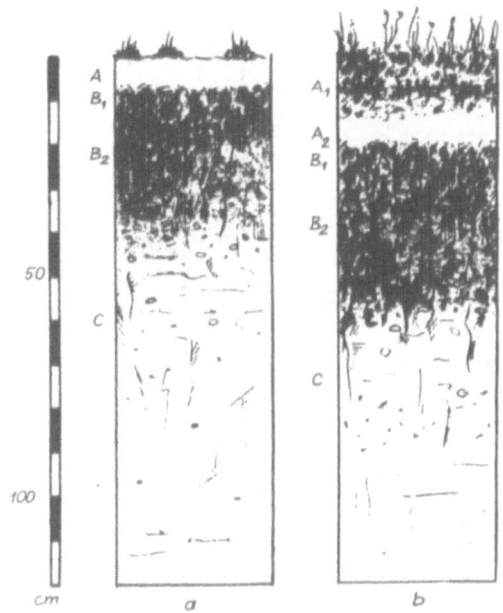

Fig. 20
Schematic profiles of solod soils

profile or if a shallow solonetz has been solodized. Due to this fact, parallel with solod formation, a relatively high ESP value may also be measured in the A horizon. Solod forming processes occur only if waterlogged and/or water saturated conditions regularly alternate with drying in the horizons involved, as is commonly the case in microdepressions.

Solod formation represented in Figure 20.a) occurs usually in heavy textured soils where, as a result of the slowness of solution movements, water saturated conditions are prolonged.

The B horizon is also often solodized and - unlike the well-developed solonetz soils - the columns or crusts of this horizon contain a considerable amount of silicon compounds and are lighter in colour.

The accumulation of sesquioxides can always be observed under the solodized horizon and, as with solonetz soils, the clay content is always higher there than in the A horizon.

Figure 20.b) represents a soil where solod formation manifests itself in the A_2 horizon. In cases like this the A_1 horizon is humous, while under the solodized A_2 horizon displaying the maximum $SiO_2 : Al_2O_3$ ratio the B horizon is also affected by solod forming processes to a greater or smaller degree.

4. Solonized and slightly salt affected soils with minor structure formation.

As mentioned before, in some cases a solonetz profile develops before the ESP value exceeds 15. A part of these soils should be considered as salt affected or solonetz soils, but their majority belongs to the potential salt affected soils.

Figure 21. demonstrates the salt profile of a solonetzic meadow soil from the eastern part of the Hungarian Lowland (Profile - Püspökladány-16).

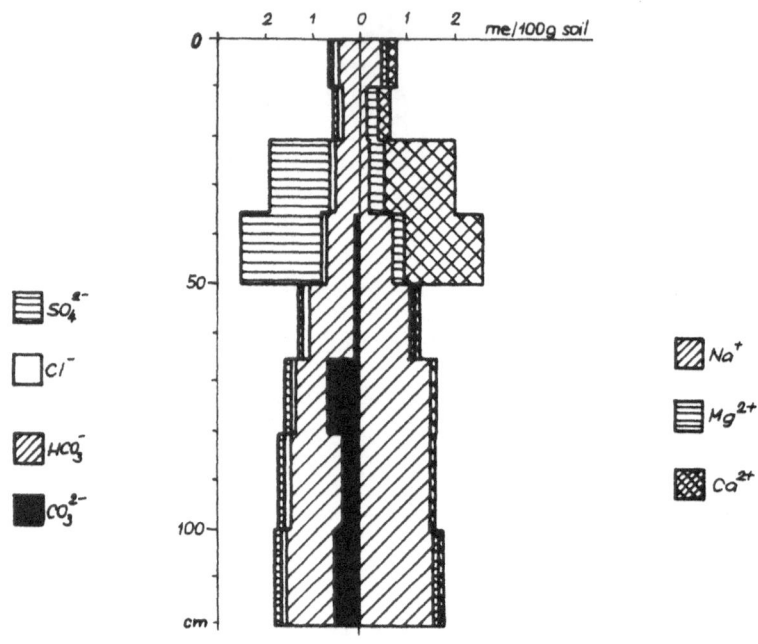

Fig. 21

Salt profile of a solonetzic meadow soil from the eastern part of
the Hungarian Lowland, Profile - Püspökladány-16., Hungary

The description of the profile of the above solonetzic meadow soil (Profile - Püspökladány-16.) reads as follows :

Surroundings : Pasture of medium quality.
Topography : Gently undulating microrelief - intermediary position.
Vegetation : Natural pasture - Festuca pseudovina, Festuca rubra, Trifolium sp.,
              Achillea millefolium, Statice Gmelini.
Depth of the profile: 130 cm.
Effervescence (with dilute acid): from 30 cm.
Alkalinity against phenolphthalein: from 40 cm.
Thickness of the humus layer: 65 cm.
Water table: 380 cm.

Genetic horizons:

A   0- 20 cm: Dark gray, dry, compact, fine granular structure, clay loam. Plentiful roots. Abrupt boundary.

B_1 20- 50 cm: Brownish black, dry, very compact, prismatic structure, silty clay. Columnar-like structure formation along the wide cracks. Moderate amount of roots. Few small iron mottles and concretions. Gradual boundary.

B_2 50- 65 cm: Yellowish gray, dry, very compact, weakly developed coarse prismatic structure, silty clay. Few roots. Many rusty mottles of iron, few lime and iron concretions. Gradual boundary.

C_1 65-100 cm: Grayish yellow, moist, moderately compact, loess-like clay loam. Rusty iron mottles, humus streaks along the roots, lime concretions (1-2 cm in diameter). Gradual boundary.

C_2 100-130 cm: Grayish yellow, moist, slightly compact, loess-like loam. Dark humus streaks, iron mottles, rusty iron spots.

Soil type: Solonetzic meadow soil on loess-like clay loam. (Salt affected soil with minor structure formation.)

As regards the formation of European solonetz and solod soils, more detailed information can be found in the publication "European Solonetz Soils and Their Reclamation" - edited by I. Szabolcs (Akadémiai Kiadó - Publishing House of the Hungarian Academy of Sciences -, Budapest, 1971.).

Potential Salt Affected Soils

Those soils are considered potential salt affected soils which are not, or only to a very low degree, saline and/or alkaline at present but human intervention, especially irrigation, may cause their considerable salinization and/or alkalinization.

It is well-known that the majority of the irrigated territories in existence in the world are exposed to the hazard of secondary salinization, alkalinization and waterlogging.

While the existing salt affected soils can be recognized on the basis of a few morphological, chemical and physico-chemical observations and determinations, the recognition of potential salt affected soils as well as of the long-term hazard of salinity or alkalinity on any given territory necessitates special survey and methods. These methods are described in detail in several books (e.g. "International Source--Book on Irrigation and Drainage of Arid Lands in Relation to Salinity and Alkalinity." Ed. by V.A. Kovda, C. van den Berg and R.M. Hagan. FAO/UNESCO 1967., "Symposium on the Reclamation of Sodic and Soda-Saline Soils" Agrokémia és Talajtan 18. Suppl. pp. 351 ff. 1969., etc.)

In the light of the paramount importance of irrigation and of the close relationship existing among irrigation, drainage and the salinity and alkalinity of soils, it was deemed necessary to delineate - whenever it is possible on the basis of the available data - those areas on the Maps of Salt Affected Soils which are exposed to the hazard of salinity or alkalinity owing to the introduction, the present practice and/or to the further extension of irrigation.

Secondary salinization and alkalinization processes may take place mainly in the following situations:

1. Accumulation of salts from irrigation water of poor quality.

2. Increase in the level of groundwater.

   a) The salt content of the groundwater accumulates in the affected layers;

   b) the rising groundwater transports the salts from the deeper soil layers to the surface or surface layers, or

   c) the rising water table limits natural drainage and hinders the leaching of salts.

The possible hazard of salinization and/or alkalinization in irrigated areas or areas to be irrigated may be determined by the following factors:

1. Climatic factors such as: temperature, rainfall, humidity, vapour pressure, evaporation and their fluctuations and dynamics;

2. Geological, geomorphological, geochemical, hydrological, hydrogeological and hydrochemical factors such as: natural drainage, the depth and fluctuation of the water table, the direction and velocity of horizontal groundwater flow, salt content and composition of the groundwater, etc.

3. Soil factors such as: soil profile, texture, structure, saturated and unsaturated water conductivity, soluble salt content, salt composition and salt profiles, exchangeable cations, pH, etc.

4. Agrotechnics such as: land use, crops, cultivation methods, etc.

5. Irrigation practices such as: the amount of irrigation water; method, frequency and intensity of irrigation; salt content and composition of irrigation water; natural and artificial drainage, etc.

The above-mentioned factors determine the aims and methods of the preliminary survey of soils in order to define the degree or the existence of potential salinity and/or alkalinity.

Evidently, the environmental conditions on the one hand, and the methods of the utilization of the territory in question on the other hand should be taken into consideration when an area is evaluated in this respect. Due to this fact different limit values and different methods - based on uniform principles - should be selected in the course of this procedure. For example in arid regions, particularly outside Europe, in deserts and semi-deserts practically all irrigated areas are potentially salt affected owing to the arid climate as well as to the high accumulation of salts in the soils and waters of these areas.

In moderate climate - where most of the lands irrigated or to be irrigated may be found in Europe -, the problems of potential salt affected soils are not the same as those encountered in arid zones because of the different climatic, geochemical and farming conditions.

In these areas the basic aims of the survey and study of potentially saline or alkaline soils are to predict the harmful processes and to elaborate, whenever it is possible, methods suitable to prevent the occurrence of secondary salinization and alkalinization.

In order to develop a reliable method of predicting salinization and alkalinization the following problems must be solved:

1. The main sources of water soluble salts (irrigation water, groundwater, surface waters, salty deep soil layers, etc.) must be identified.

2. The main features of the salt regime must be characterized (salt balances); and the whole range of natural factors influencing the salt regime must be analysed.

3. The effect of irrigation on the water and salt regimes of the soil must be determined.

Consequently, an exact salinity and/or alkalinity prognosis must be based on an evaluation of many natural and human factors and a knowledge of the existing soil processes.

Based on these principles (see details in: Symposium on the Reclamation of Sodic and Soda-Saline Soils. Agrokémia és Talajtan, 18. Suppl. pp. 351-376. 1969.) a special survey was made in the eastern part of the Hungarian Lowland in order to predict the influence of existing and projected irrigations on soil salinity and alkalinity.

In Figure 22. an overall map of the eastern part of the Hungarian Lowland may be seen, showing with schematic indication: 1. the areas to be irrigated, where practically no hazard of salinity and alkalinity exists; 2. the areas which may be irrigated conditionally because the hazard of salinity and alkalinity is considerable if irrigation is extended; in case of irrigation the precise and permanent control of the salt balance of soils is necessary in these areas; 3. the areas not to be irrigated because any irrigation would cause immediately secondary salinization and/or alkalinization.

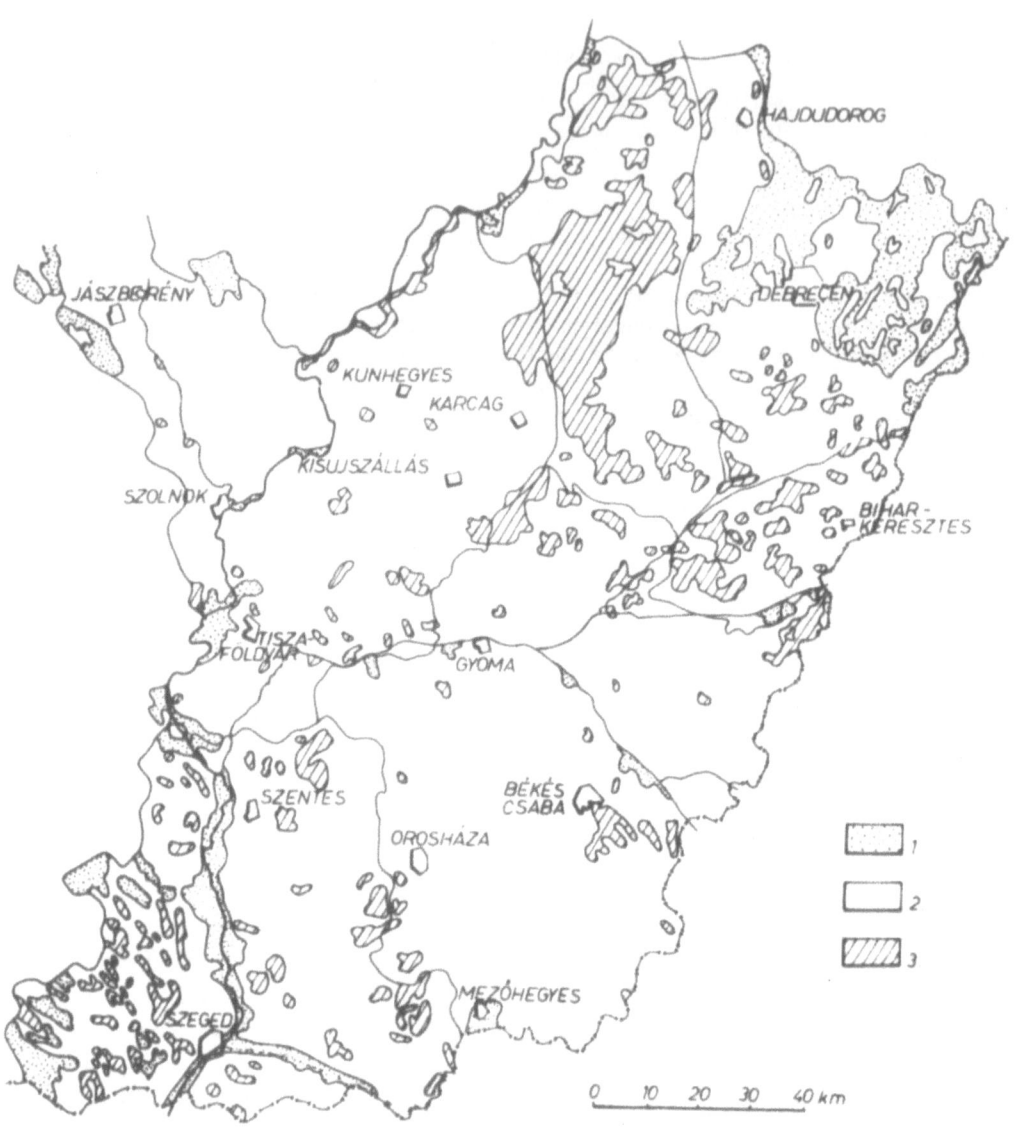

Fig. 22
The overall map of the eastern part of the Hungarian Lowland,
indicating the general possibilities of irrigation. 1. Areas where
irrigation does not involve the hazard of salinity and alkalinity.
2. Areas which may be irrigated conditionally. 3. Areas not to
be irrigated.

Correlation of the Classification of Salt Affected Soils Elaborated by the Subcommission on Salt Affected Soils of the ISSS with Other Classification Systems

Considering that numerous classification and soil grouping systems exist, any such system, elaborated for international adaptation and use, must necessarily correlate the different soil denominations.

Most of the national soil classification systems used in European countries may be easily correlated with the one adopted by the Subcommission. The correlation can be elaborated for soil maps on different scales.

In Table 1. the correlation between the Hungarian classification system (used in the preparation of the Map of Salt Affected Soils of Hungary, on scale 1 : 500,000) and the one developed by the Subcommission is given.

In Table 2.- based on the Tentative Correlation Table prepared by R. Dudal, Chief of the Soil Resources, Development and Conservation Service; Land and Water Development Division of FAO - the tentative correlation of the most widely used classifications is given. Table 2.shows that the correlations are quite satisfactory for general maps from the practical point of view.

Table 1

CORRELATION BETWEEN THE HUNGARIAN CLASSIFICATION SYSTEM AND THE ONE DEVELOPED BY THE SUBCOMMISSION ON SALT AFFECTED SOILS OF THE ISSS

Classification units according to the Hungarian soil classification	Classification units according to the classification accepted by the Subcommission on Salt Affected Soils of the ISSS
Chloride and/or sulphate solonchak	Saline soils
Soda solonchak Soda solonchak-solonetz	Alkali soils without structural B horizon
Calcareous meadow solonetz Calcareous meadow solonetz turning into steppe formation Calcareous solonetzic meadow soil	Alkali soils with structural B horizon (calcareous)
Meadow solonetz Meadow solonetz turning into steppe formation Solonetzic meadow soil	Alkali soils with structural B horizon (non-calcareous)
Chernozems and meadow chernozems salty in deeper layers Potential salt affected soils	Potential salt affected soils

Table 2

TENTATIVE CORRELATION OF THE MOST WIDELY USED CLASSIFICATION SYSTEMS OF SALT AFFECTED SOILS

Subcommission on Salt Affected Soils Classification			Soil Map of the World (Soil Map of Europe) FAO/UNESCO Project (ECA Working Party) Classification	Australian Classification 1968	Canadian Classification 1965	French Classification 1967	USDA Classification 1967	USSR Classification 1967
Basic Grouping 1967	In the legend of the Map of European Salt Affected Soils 1968	Saline and Sodic Soils Map of Australia 1971						
SALINE SOILS	SALINE SOILS	SALINE SOILS	SOLONCHAK Orthic solonchak	Solonchak		Sols salins (excepté sous-groupe acidifié) Sols salins à alcalins	Salorthids	Fluffy solonchak (non steppic) Crust solonchak Soda solonchak (non steppic)
		ALKALINE SODIC SOILS AS1 AS2 AS3	Mollic solonchak		Saline subgroups (pp)		Salorthidic Calciustolls Salorthidic Haplustolls	Fluffy solonchak (steppic) Soda solonchak (steppic) Takyrs
ALKALI SOILS without structural B horizon	ALKALI SOILS without structural B horizon		Takyric solonchak					
			Gleyic solonchak			Sols à gley salés	Halaquepts (pp)	Meadow solonchak
ALKALI SOILS with structural B horizon	ALKALI SOILS with structural B horizon Solonchak-solonetz and calcareous solonetz	NON-ALKALINE SODIC SOILS NS1 NS2	SOLONETZ	Solonetz and solodized solonetz	Brown solonetz (pp) and alkali solonetz	Sols sodiques à horizon B et solonetz solodisés	Nadurargids Natrargids Natriboralfs Natrudalfs Natrustalfs Natrixeralfs	Desert-steppe and Desert solonetz
	Non-calcareous solonetz with A horizon < 15 cm Solodized and/or deeply leached solonetz		Orthic solonetz					
	Solonized and slightly salt affected soils with minor structure formation		Mollic solonetz		Black, gray and brown solonetz (pp)		Natralbolls Natriborolls Natrustolls Natrixerolls	Steppe solonetz
			Gleyic solonetz		Gleyed solonetz		Natraquolls Natraqualfs	Meadow solonetz
	Deeply leached solodized soils and solods		Solodic planosols	Solod and solodic groups (pp)	Solod and solodic subgroups (pp)	Solods	Argiabolls (pp)	Solod

* (pp): in part

40

The Extent of Salt Affected Soils in Europe

Salt affected soils occur in considerable extent in at least twelve European countries, while in some others they may be found only in very small spots. The distribution of salt affected soils in Europe, just like on other continents, is in close relationship with the climatic, geological and, particularly, hydrogeological conditions. The most part of European salt affected soils may be found in the semi-arid, steppe and forest steppe regions of the USSR, on the second and third lowlands of the Danube river (on the territories of Czechoslovakia, Hungary, Yugoslavia and Rumania) and in Spain. Nevertheless, as far as annual temperature, precipitation and local altitude are concerned, the conditions of the occurrence of salt affected soils may be quite different. In Table 3. the regions of occurrence of salt affected soils in twelve European countries are given, with the indication of some climatic characteristics and of altitude.

In Table 4. the data showing the extent of the different types of salt affected soils, as well as their percentage of the total salt affected area are given for twelve European countries, based on information conveyed by the contributors. The data available at present are not sufficient for the accurate, numerical characterization of every classification unit in every country. This is the reason why no numerical estimate relating to the whole continent has been made from the data presented in Table 4.

Country, region	Mean annual temperature °C	Mean annual precipitation mm	Altitude m
AUSTRIA			
Pulkantal (Illmitz)	9.2	566	187
Seewinkel (Apetlon)	9.8	623	140
BULGARIA			
Danube Valley	11.6	585	50-100
Maritza, Tundja, Strema Valley	12.3	500	100-200
CZECHOSLOVAKIA			
South Moravia	9.0	500-550	200
Danubian Lowland	9.0-10.0	500-600	100-150
East Slovakian Lowland	9.0	600-650	100
FRANCE			
Atlantic sea coast	10.5-11.5	700-800	0-100
Mediterranean sea coast	14.0-15.0	550-650	0-100
GREECE			
Ionian sea coast	17.5-18.5	650-750	0-100
Thessaloniki Plain	14.0-15.0	500-600	0-50
HUNGARY			
Hungarian Plain	10.0-10.5	524-585	80-120
ITALY			
Northern Italy	13.7	744	0-100
Southern and island region	17.6	478	0-100

PORTUGAL			
Atlantic sea coast	14.0-18.0	660-1400	0-100
RUMANIA			
Black Sea shore - Danube Delta	11.0-11.3	359- 439	0- 50
NE Romanian Lower Danube Plain	9.6-11.1	400- 515	0-100
Western Romanian Lower Danube Plain	10.6-11.5	480- 570	100-200
Tisza Plain	10.7-10.8	558- 620	80-100
Moldavian Table-Land	9.0-10.5	399- 588	200
SPAIN			
Southern region	17.6-18.3	535- 651	30-100
South-Western region	17.0-17.6	295- 419	0- 60
Basin depression of the Ebro	14.6-15.1	324- 378	118-380
Central plateaus	11.6-14.2	389- 403	700
USSR			
Ukraine	7.0	500- 550	100
Oka-Don Plain	7.0- 8.0	500- 550	100-200
Preazovian Plain	9.0-10.0	300- 550	100
Prevolgian Plateau	6.5- 7.5	500- 600	200-300
Precaspian Lowland	6.0	150- 250	0-100
Transcaucasian plains	14.0	200- 250	100-200
Armenia	11.3-14.0	200- 220	500-700
Transvolga region	4.0- 5.0	300- 400	100-200
YUGOSLAVIA			
Vojvodina	10.0-10.5	550- 600	100
Macedonia	11.2	652	100-200
Adriatic sea coast	14.5-15.5	500- 900	0- 50

Table 4

DISTRIBUTION AND EXTENT OF SALT AFFECTED SOILS IN EUROPE

Countries and originators	Mapping units (area in 1000 ha/area in percentage of the total salt affected area)					Total area in 1000 ha.
	Saline soil	Alkali soil without structural B horizon	Alkali soil with structural B horizon		Potential salt affected soil	
			non-calc.	calc.		
AUSTRIA J. Fink O. Nestroy	0.5 / 25.0	- / -	- / -	- / -	2.5 / 75.0	3.0
BULGARIA L. P. Raikov H. Trashliev	5.0 / 20.0	- / -	20.0 / 80.0	- / -	- / -	25.0
CZECHOSLOVAKIA J. Hrasko	6.2 / 5.8	7.5 / 7.1	2.7 / 2.5	4.3 / 4.1	85.0 / 80.5	105.7
FRANCE E. Servat J. Servant	175.0 / 70.0	- / -	75.0 / 30.0	- / -	- / -	250.0
GREECE E. P. Papanicolaou	/	/	/	/	/	3.5

HUNGARY						
I. Szabolcs	1.6 / 0.1	58.6 / 4.7	294.0 / 23.1	31.9 / 2.5	885.5 / 69.6	1271.6
G. Várallyay						
J. Mélyvölgyi						
ITALY						
O.T. Rotini	50.0 / 20.0	– / –	– / –	– / –	400.0 / 80.0	450.0
L. Carloni						
PORTUGAL						
J. Carvalho Cardoso	/	/	/	/	/	25.0
M. Branco Marado						
RUMANIA						
N. Florea	40.0 / 16.0	100.0 / 40.0	110.0 / 44.0		–	250.0
J. Munteanu						
SPAIN						
P. Grajera Torrez	/	/	/	/	/	840.0
R. Grande Covian						
J. Bardaji Cando						
J.M. Ontanon						
USSR						
V.V. Egorov	7546.0 / 16.0	1616.0 / 3.4	20382.0 / 43.1	– / –	17781.0 / 37.5	47325.0
N.I. Bazilevich						
E.I. Pankova						
YUGOSLAVIA						
N. Miljkovic	20.0 / 7.8	50.0 / 19.6	110.0 / 43.1	75.0 / 29.5	– / –	255.0
G. Filipovski						
N. Plamenac						
P. Blaskovic						

Acknowledgements

The Department of Environmental Sciences and Natural Resource Research (formerly Department of Advancement of Sciences, Division of Natural Resources) of UNESCO has lent much appreciated permanent support and financial assistance to the World Map of Salt Affected Soils Project in general, and to the Map of European Salt Affected Soils in particular. Special thanks go to Dr. M. Batisse, Director of the Department, as well as to Dr. S. Evteev and Dr. K.O. Lange, Programme Specialists, for having effectively contributed to the substantial progress of the Project.

The counsel and assistance of Dr. R. Dudal, Chief, Soil Resources, Development and Conservation Service, Land and Water Development Division, and, previously, of Dr. L.D. Swindale, FAO, are also highly valued.

Most deeply felt gratitude is due to Prof. V.A. Kovda, both in his capacity as former Director of the Department of Advancement of Sciences, UNESCO, and President of the International Society of Soil Science and to Prof. F.A. van Baren, Secretary General of the International Society of Soil Science, who have displayed an extraordinary interest in the Project. Without their invaluable support enjoyed throughout the whole process of work this Project could not have been realized.

Special mention is to be made of the Hungarian Academy of Sciences which has guaranteed the proper conditions and provided financial assistance for our activities centered at the Research Institute for Soil Science and Agricultural Chemistry, as well as of the Hungarian National Commission for UNESCO for kindly sponsoring this work.

Acknowledgement is also due to CARTOGRAPHIA, Budapest, for their assistance in the cartographical work and for undertaking the printing of maps.

Numerous experts and colleagues, particularly the staff-members of the Research Institute for Soil Science and Agricultural Chemistry have contributed to the compilation of maps and related materials; their help and efforts are highly appreciated. The Editor wishes to express his thanks also to Mrs. K. Keszthelyi for her assistance in the preparation of this publication.

Bibliography

Compiled by G. Várallyay

ABELL, L.F. - GELDERMAN, W.J.: Annotated bibliography on reclamation and improvement of saline and alkali soils (1957-1964). (E) Int. Inst. Land Recl. Impr., Wageningen. Bibl. No.4., 1964.

ÁBRAHÁM, L.: (Low rates of ameliorants for the improvement of salt-affected soils /Hungary/). (R,e) Agrokém. Talajt. 14 Suppl. 329-332, 1965.

ÁBRAHÁM, L.: (Problems of utilization and amelioration of solonchak-solonetz soils in the Danube valley /Hungary/). (H,e,f,r) Agrokém. Talajt. 17 283-296, 1968.

ÁBRAHÁM, L.: (Gypsum as soil ameliorant /Hungary/). (H) Agrokém. Talajt. 19 173-192, 1970.

ÁBRAHÁM, L. - BOCSKAI, J.: (Amelioration of Hungarian salt-affected soils.) (H,r) OMMI Publ., Ser.1, No.4, Budapest, 1971.

ÁBRAHÁM, L. - BOCSKAI, J.: The utilization and amelioration of solonetz soils in Hungary. (E) In: I. Szabolcs: "European solonetz soils and their reclamation". 61--97. Akadémiai Kiadó, Budapest, 1971.

ÁBRAHÁM, L. - SZABOLCS, I.: Improving alkali soils with small doses of reclamation materials. /Hungary/. (E,f,g) Trans. 8th Int. Congr. Soil Sci., 2 875-880. 1964.

ADERIKIN, P.G.: (Formation of salt-affekted soils in the Kamennoi steppe region / USSR/). (R) Pochvovedenie, (6) 63-66, 1953.

AFANASEV, G.V. - KASANSKY, A.D.: (Salt-affected soils and the formation of solonchaks in the northern Dvina delta area /USSR/). (R) Dokl. mosk. sel.-khoz. Akad. K.A. Timiryazeva, 84 74-78, 1963.

AGABABJAN, V.G.: (Soda solonchaks in the Ararat Plain and their amelioration with acids /USSR/). (R) Tretij delegatsk. s'ezd pocsv., 207-201. Izd. Nauka Moskva, 1968.

ALEXIADES, C.A.: (Contribution to the study of the great soil groups of Greece /terra rossa, rendzina, red loam, solonchak, solonetz/). (Gr,e) Aristot. Panepist. Thessalonikis Epet. Geopon. Dasolog. Skol., 164-200, 1967.

ALPHEN, J.G. van: Some notes on the reclamation of sodic soils. (E) Ann. Rep. Int. Inst. Land Recl. Impr., Wageningen, 28-36, 1966.

ALPHEN, J.G. van - ABELL, L.F.: Annotated bibliography on reclamation and improvement of saline and alkali soils (1960-1966). (E) Int. Inst. Land Recl. Impr., Wageningen, Bibl. No. 6., 1967.

ALVIM, A.J.S.: (Amelioration of "sapais" /Portugal/). (Port.) Pedologia, 3 245-261, 1968.

ALVIM, A.J.S. - VEIGUINHA, A.S.: (Amelioration of salt-marsh soils /Portugal/).
(Port., e) Agronomia Lusit., 25 1133-1156, 1963.

ANANYAN, A.K.: (Drainage for the amelioration of soda solonchaks /USSR). (R)
Izd. Kolos, Moskva, 1971.

ANTIPOV-KARATAEV, I.N.: (Amelioration of solonetzes in the Soviet Union). (R) Izd.
Ak. Nauk SSSR, Moskva, 1953.

ANTIPOV-KARATAEV, I.N.: (Physico-chemical studies in connection with the amelioration of solonetzes (USSR). G,e,f) Trans. 7th Int. Congr. Soil Sci. 1 571-576,
1960.

ANTIPOV-KARATAEV, I.N. - KADER, G.M.: (Sodic solonetzes, their genesis and
amelioration in the Soviet Union). (R,e) Agrokém. Talajt. 14 Suppl. 111-114, 1965.

ANTIPOV-KARATAEV, I.N. - PAK, K.P.: (Solonetzes and their amelioration under
irrigated and non-irrigated conditions /USSR/. (R) Pochvovedenie, (10) 1-6, 1965.

ARANY, S.: (Salt-affected soils and their reclamation (Hungary/). (H) Mezőgazd. Kiadó, Budapest, 1956.

ARANY, S.: (Classification of the Hungarian salt-affected soils.) (G,r,e) A. Thaer
Arch. 4 23-36, 1960.

ARANY, S.: (History of the "digo"-method for the amelioration of salt-affected soils
in Hungary /literature review/). (H) Agrokém. Talajt. 13 361-369, 1964.

AUBERT, G.: Arid zone soils. A study of their formation, characteristics,utilization
and conservation. (E) Arid Zone Res., 18 (The problems of the arid zone. Proc.
Paris Symp.) 115-138, 1962.

AXENOVA, I. - MAIANU, A. - ALBESCU, I.: Dynamics of soil nutrients, salts and
rice production during the reclamation of a secondary saline soil in the Danube
floodplain /Rumania/. (E) Stiinta Solului, 6 (2-3) 34-39, 1968.

AYERS, A.D.: Improvement of saline and sodic soils. (E,f,g,sp) Potassium Symp.,
Athens, 259-270, 1962.

AYERS, A.D. et al.: Saline and sodic soils of Spain. (E) Soil Sci. 90 133-138, 1960.

BACSÓ, A. - FEKETE, J.: Role of groundwater in the secondary salinization of meadow chernozem soils in the Hajduság area /Hungary/. (E) Agrokém. Talajt. 18 Suppl.
339-340, 1969.

BALLENEGGER, R.: (Salt affected soils and their amelioration /Hungary/). (H) Egye-
temi Nyomda, Budapest, 1931.

BARDAJI, J. - RISSEEW, I.: (Amelioration of saline-alkali soils in the Guadalquivir
delta area /Spain/). (Sp) IRYDA información. Salt Affected Soils Subcommission
Meeting, Sevilla, 1971, 67-76.

BAROCCIO, A.: (Secondary effects of gypsum application /Italy/). (I) Ann. Staz. chim.
-agr. Roma, 3 (185) 8-16, 1961.

BARRADAS, F.H.: (Sand for the reclamation of "sapais" /Portugal/). (Port)Pedologia,
2 37-49, 1967.

BAZILEVICH, N.I.: (Geochemistry of sodic soils). (R) Izd. Nauka, Moskva, 1965.

BAZILEVICH, N.I.: (Forest steppe solods /USSR/. (R) Izd. Nauka, Moskva, 1967.

BAZILEVICH, N.I. - PANKOVA, E.I.: (Tentative classification of soils according to salinity /USSR/).(R, e) Pochvovedenie, (11) 3-16, 1968.

BAZILEVICH, N.I. - PANKOVA, E.I.: Classification of soils according to their chemistry and degree of salinization /USSR/. (E) Agrokém. Talajt. 18 Suppl. 219-226, 1969.

BAZILEVICH, N.I. - PANKOVA, E.I.: (Characteristics of salt affected soils). (R) In: Egorov, V.V. - Bazilevich, N.I.: "Salt affected soils in the European part of USSR and the Transcaucasian Soviet Republics", 21-203. Nauchn. Trudy pochv. Inst. Dokuchaeva, Moskva, 1973.

BAZILEVICH, N.I. - ROZOV, N.N.: (Distribution of salt affected soils with various salt compositions /USSR/). (R) In: Egorov, V.V. - Bazilevich, N.I.: "Salt affected soils in the European part of USSR and the Transcaucasian Soviet Republics", 14--20. Nauchn. Trudy pochv. Inst. Dokuchaeva, Moskva, 1973.

BAZILEVICH, N.I. - KOZLOVSKY, F.I. - PANKOVA, E.I.: (Physico-geographical factors in salt affected areas /USSR/). (R) In: Egorov, V.V. - Bazilevich, N.I.: "Salt affected soils in the European part of USSR and the Transcaucasian Soviet Republics", 8-13. Nauchn. Trudy pocsv Inst. Dokuchaeva, Moskva, 1973.

BIRNYUKOVA, A.P.: (Effect of irrigation on the water and salt regimes of soils in the southern Transvolga region /USSR/). (R) Izd. Akad. Nauk SSSR, Moskva, 1962.

BOCSKAI, J.: (Effect of various agrotechnical factors on crop yields on a solonetz soil /Hungary/). (H, r, e, g) Talajtermékenység, 3 (1) 61-77, 1968.

BOCSKAI, J.: The use of acid resins, supplied by the oil industry, in the chemical amelioration of alkali soils /Hungary/. (E) Agrokém. Talajt. 18 Suppl. 336-338. 1969.

BOL'SHAKOV, A.F. et al.: (Changes in soil forming processes in solonchakous solonetzes due to amelioration /USSR/). (R, e) Pochvovedenie, (6) 11-22, 1966.

BOLYSHEV, N.N.: (Characteristics and composition of the adsorption complex of solonetz soils /USSR/). (R, e) Agrokém. Talajt. 14 Suppl. 359-363, 1965.

BOROVSKY, V.M. - POGREBINSKY, M.A.: (Soil formation and soil amelioration in continental deltas /USSR/). (R, e) Plod. melior. pochv. SSSR (Dokl. VIII. Mezhd. Kongr. Pochv.) 173-181, 1964.

BOUMANS, J.H. - MOLEN, W.H. van der: (Drainage requirements of irrigated soils in relation to their salinity). (Dutch, e) Landb. Tijdschr., 76 880-887, 1964.

BOWER, C.A. - FIREMAN, M.: Saline and alkali soils. (E) Yearbook of Agriculture, 282-290. US Dept. Agric. Washington, 1957.

BUHOCI, L. - IONESCU, C. - COVALIOV, T.: (Soil and groundwater salinization in the dammed enclosure from Calmatui-Gropeni /Rumania/). (Rm, e, f, r) Studii Hidroamelior., 4 263-281, 1968.

CARVALHO CARDOSO, J.: Reclamation of saline soils in Portugal. (E) Agrokém. Talajt. 18 Suppl. 329-335, 1969.

CERVENKA, L.: (Origin of salt-affected soils in the southern part of Slovakia). (Cz, r, e) Vedecke práce VUZH, Bratislava, 222-247, 1961.

CERVENKA, L.: (Salt-affected soils in the southern Moravia /Slovakia/). (Cz, r, e) Vedecke práce Labor. Podozn, 3, 1968.

CERVENKA, L. - LOPATNIK, J.: (Salt-affected soils in Slovakia). (Cz,r,g) Rostl. Vyroba, 33 1383-1398, 1960.

CERVENKA, L. - SZABOLCS, I.: (Genesis and properties of Slovakian salt-affected soils). (H,e,g,r) Agrokém. Talajt. 17 269-282, 1968.

CHIKVISHVILI, V.I.: (Solonetz soils in Georgia, their amelioration and agricultural utilization). (R) Trudy Inst. Pochvoved. Gruz. SSR, T. II., Tbilisi, 1963.

CHIKVISHVILI, V.I.: (Salt affected soils in Georgia, their amelioration and agricultural utilization). (R) Sborn. stat. VIII. Mezhd. Kongr. Pochv., Tbilisi, 1964.

DARAB, K.: (Salt balance and salt regime of irrigated soils in Hungary). (H,r,e) Agrokém. Talajt. 10 305-314, 1961.

DARAB, K.: (The application of soil genetic principles to irrigation in the Hungarian plain). (H,r,e,g,f) OMMI Publ. Ser. 1, No.4, Budapest, 1962.

DARAB, K.: Chemical and physico-chemical effects of sodium carbonate in soils. (E,r) Agrokém. Talajt. 14 Suppl. 175-183, 1965.

DARAB, K.: (Remarks on the paper by Dr. H.Franz entitled: "Data to the quaternary stratification and to the genesis of salt-affected soils in the Hortobágy region and in its surroundings /Hungary/). (H,g,e,r) Agrokém. Talajt. 16 459-476, 1967.

DARAB, K. - FERENCZ, K.: (Soil mapping and control of irrigated areas /Hungary/). (H,r,e) OMMI Publ., Ser.1, No.10, Budapest, 1969.

DARAB, K. - SZABOLCS, I.: Types of secondary alkalinization of soils in irrigated zones of the region of the Great Hungarian Plain (E,f,g) Trans. 7th Int. Congr.Soil Sci., 1 535-542, 1960.

DIMO, N.A.: (Types of soil salinization /USSR/). (R) Izd. Kievsk. Univ., Kiev, 1960.

DOKUCHAEV, V.V.: (Our steppes before and now /Russia/). (R) 1892. Izbr. Soch., 449-512. Sel'khozgiz, Moskva, 1954.

DONCHEV, I. - TRASHLIEV, Ch.: (The formation of soda and the solonetzization of some Bulgarian meadow soils). (B,r,e) Nauchn. Trudy nauchno-issled. Inst. Pochv. "N. Pushkarov", 3 195-227, 1957.

(Drainage in the amelioration of salt affected soils /USSR/). (R) Izd. Akad. Nauk SSSR, Moskva, 1958.

DUDAL, R.: Definitions of soil units for the soil map of the world. (E) World Soil Res. Rep. No. 33., 1-72., FAO, Rome, 1968.

DUDAL, R.: Supplement to definitions of soil units for the soil map of the world. (E) World Soil Res. Rep. No. 37., 1-10. FAO, Rome, 1969.

DUDAL, R.: Key to soil units for the soil map of the world. (E) Soil Map of the World FAO/UNESCO Project. FAO, Rome, 1970.

DUDAL, R. - TAVERNIER, R. - OSMOND, D.: Soil Map of Europe. 1:2,500,000. (E) FAO, Rome, 1966.

EGOROV, V.V.: (Sodic soils in the Soviet Union and methods of their amelioration). (R,e) Agrokém. Talajt. 14 99-106. 1965.

EGOROV, V.V.: Some aspects relating to the sodic salinization in the subhumid regions of Europe and Asia. (E) Agrokém. Talajt. 18 Suppl. 187-191, 1969.

EGOROV, V.V. - BAZILEVICH, N.I.: (Salt affected soils in the European part of USSR and the Transcaucasian Soviet Republics). (R) Nauchn. Trudy pochv. Inst.Dokuchaeva, Moskva, 1973.

EGOROV, V.V. - BAZILEVICH, N.I. - PANKOVA, E.I.: (Construction principles of the map of salt affected soils' types /USSR/). (R) In: Egorov, V.V. - Bazilevich, N.I. "Salt affected soils in the European part of USSR and the Transcaucasian Soviet Republics". 4-7. Nauchn. Trudy pochv. Inst. Dokuchaeva, Moskva, 1973.

EGOROV, V.V. - POPOV, A.A. - KONOVALOV, N.N.: (Extended soil-ameliorative mapping of the Volga-Akhtuba bottomland /USSR/). (R) Pochvovedenie, (3) 16-29, 1962.

EGOROV, V.V. - POPOV, A.A. - KONOVALOV, N.N.: (Schematic soil-ameliorative zoning of the Volga delta /USSR/). (R, e) Pochvovedenie, (9) 4-13, 1962.

EGOROV, V.V. et al.: Problems in the reclamation of some saline soils in the Soviet Union. (E, f, g) Trans. 8th Int. Congr. Soil Sci., 2 903-908, 1964.

FILIPOVSKI, G.: (Genesis, evolution and scientific basis for the reclamation of salt-affected soils in the Ovce Pole /Yugoslavia/). (Sr-Cr) Ann. Book Fac. Agric. Forestry Univ. Skopje, (12) 1960.

FILIPOVSKI, G.: Results of investigations on the possibilities of reclaiming salt-affected sodic soils in Yugoslavia. (E) Agrokém. Talajt. 18 Suppl. 385-389. 1969.

FLOREA, N.: (Interpretation of the results of chemical analysis of ground-water for soil genesis and amelioration /Rumania/). (Rm, r, f) Probl. Agric., 13 (7) 1961.

FLOREA, N.: (Interpretation of the results of chemical analysis of salt-affected soils for genesis and amelioration /Rumania/). (Rm, r, f) Probl. Agric., 13 (10) 1961.

FLOREA, N. - MUNTEANU, I.: Soda saline soils in Rumania. (E) Agrokém. Talajt. 18 Suppl. 207-218, 1969.

FLOREA, N. - STOICA, L. - IVANOV, N.: (The accumulation of salts within the soils of the north-eastern Romanian plain /area between the rivers Buzau and Calmatui/). (Rm, r, f, e) Studii Pedol. Ser.C., (11) 163-193, 1963.

FRANZ, H. - HUSZ, G.: (Saline soils and the age of the saline steppes in the "Seewinkel" region /Austria/). (G) Mitt. Österr. Bodenk. Ges., (6) 138-174, 1961.

FRIDLAND, V.M. - KARAEVA, Z.S.: (Origin of acidic salt-affected soils /USSR/). (R) Pochvovedenie, (7) 77-81, 1962.

GEDROIZ, K.K.: (Solodization of soils /USSR/). (R) Izd. Nosovskaya opyt. st., 44. Leningrad, 1926.

GEDROIZ, K.K.: (Solonetz soils, their origin, properties and amelioration /USSR/). (R) Izd. Nosovskaya opyt. st., 46 3-73. Leningrad, 1928.
(Genesis and classification of semidesert soils /USSR/). (R) Izd. Akad. Nauk SSSR, Moskva, 1966.
(Genesis, regime and amelioration of salt affected soils /USSR/). (R) Trudy pochv. Inst. Dokuchaeva, 54 1-302. Izd. Akad. Nauk SSSR, Moskva, 1958.

GERASIMOV, I.P.: (Modern Dokuchaevian-conception of soil classification and its use for the preparation of the soil map of USSR and for the World Soil Map). (R) Pochvovedenie, (6) 1-14, 1964.

GERASIMOV, I.P. - GLAZOVSKAYA, M.A.: (Fundamentals of soil science and soil geography). (R) Gos. Izd. Geogr. Lit., Moskva, 1960.

GERASIMOV, I.P. - IVANOVA, E.N.: (Three scientific concepts in soil classification and their relationships). (R) Pochvovedenie, (11) 1-18), 1958.

GERASIMOV, I.P. - IVANOVA, E.N.: (Soils of Central Europe and their physico-geographical aspects). (R) Izd. Akad. Nauk SSSR, Moskva, 1960.

GEREI, L.: Transformation and destruction of clay minerals in alkali soils as affected by soil forming processes. /Hungary/. (E) Agrokém. Talajt. $\underline{17}$ 119-124. Suppl. 1968.

GLAZOVSKAYA, M.A.: (Principles of world soil classification). (R) Pochvovedenie, (8) 1-22, 1966.

GLINKA, K.D.: (Soil science). (R) Sel'khozgiz, Moskva, 1931.

GLINKA, K.D.: (Types of soil formation, their classification and geographical distribution). (G) Borntraeger, Berlin, 1914.

GORBUNOV, N.I.: (Sodium, potassium and phosphorus reserves in solonetzes in relation to their mineralogical composition and particle size /USSR/). (R, e) Pochvovedenie, (5) 67-80, 1969.

GOUNY, P. - CABIBEL, B. - LECOCQ, A.: (Sodic soils in the Vancluse region /France/). (F) C.r. hebd. Séances Acad. Agric. Fr., $\underline{54}$ 923-928, 1968.

GRANDE COVIAN, R.: Reclaiming the Guadalquivir marshes /Spain/. (E) Fatis Review, $\underline{12}$ (3) 73-76, 1965.

GRANDE COVIAN, R.: (Saline soils in delta areas of southern Spain. Evalution of saline soils in the Guadalquivir delta). (Sp) Agrokém. Talajt. $\underline{17}$ Suppl. 113-118, 1968.

GRANDE COVIAN, R.: Salt affected soils in the Guadalquivir River delta (Sevilla, Spain). (E) In: I. Szabolcs: "European solonetz soils and their reclamation". 131--149. Akadémiai Kiadó, Budapest, 1971.

GREENE, H.: Using salty lands. (E) FAO Agric. Studies, $\underline{3}$ Rome, 1948.

GRIN, G.S.: (Salt affected soils in the Ukraine and their origin). (R) Trudy kharkov. sel'khoz. Inst., $\underline{39}$ 8-102, 1962.

GRINCHENKO, A.M. (Editor): (Origin and amelioration of solonetz soils in the Ukraine). (R) Trudy kharkov. sel'khoz. Inst., $\underline{39}$ 1962.

GUSTIUC, L. - GHEORGIU, E. - POENESCU, M.: (Saline soils, having fluviomaritime origin, in the Danube delta region /Rumania/). (Rm, r, f) Lucr. st. Agr. "Prof. Ion Ionescu de la Brad", Iasi, 367-388, 1962.

GYUROV, G.: (Peaty and alluvial meadow soils in the salt affected soil region beside the Marica river /Bulgaria/). (B, r, g) Nauchn. Trud. Vish. selskostop. Inst."Vasil Kolarov", $\underline{13}$ (1) 211-221, 1964.

GYUROV, G.: (Magnesium solonetzization of chernozem-smonitzas in Bulgaria). (B, r, g) Nauchn. Trud. Vish. selskostop. Inst. "Vasil Kolarov", $\underline{15}$ (1) 195-203, 1966.

HALKIAS, N.A.: Experimental work on the reclamation of saline and alkali soils in the Thessaloniki plain, Greece. (E) Greek Nat. Comm. Irrig. Drain., Athens, 1965.

HARMATI, I.: (Amelioration and irrigation of salt affected soils /Hungary/). MTA Agrártud. Oszt. Közlem. $\underline{25}$ 26-33, 1966.

HERKE, S.: (The role of hydrological conditions in the genesis of salt affected soils in the region between the rivers Danube and Tisza /Hungary/ and on changes in their characteristics.) (G,e,f,r) Agrokém. Talajt. 13 Suppl. 157-164, 1964.

HERKE, S. - HARMATI, I.: Amelioration and utilization of alkali soils of the solonchak and solonchak-solonetz types in the region between the rivers Danube and Tisza /Hungary/. (E,r) Agrokém. Talajt. 14 Suppl. 313-322, 1965.

HILGARD, E.W.: Soils, their formation, properties, composition and relations to climate and plant growth on the humis and arid regions. (E) MacMillan, New York, 1906.

HISSINK, D.J.: Base exchange in soils (E) Trans. Farad. Soc., XX (60), 1925.

HRASKO, J.: Salt sources of alkali soils in southern Slovakia. (E) Agrokém. Talajt. 17 Suppl. 105-112, 1968.

HRASKO, J.: Salt affected soils in Czechoslovakia and the problems of their utilization. (E) In: I. Szabolcs: "European solonetz soils and their reclamation". 49-60. Akadémiai Kiadó, Budapest, 1971.

HRASKO, J. - CERVENKA, L.: (Sodic soils in Czechoslovakia). (R,e) Agrokém. Talajt. 14 Suppl. 391-400, 1965.
(The Hungarian salt affected soils). (H) Földmüv. Min. kiadv., Budapest, 1934.

HUSZ, G.: (Theory and practice of the amelioration of salt affected soils with special regard to conditions in the "Seewinkel" region /Austria/. I. Theory). (G) Bodenkultur, 16 223-244, 1965.

HUSZ, G.: (Theory and practice of the amelioration of salt affected soils with special regard to conditions in the "Seewinkel" region /Austria/. II. Experiments, results and practical conclusions). (G,e) Bodenkultur, 17 1-34, 1966.

HUSZ, G.: (Systematics of salt affected soils in the "Seewinkel" region /Austria/). (G,e) Bodenkultur, 17 295-309, 1966.

HUSZ, G.: Contribution to the discussion about the classification and cartography of salt affected soils. (E) Agrokém. Talajt. 18 Suppl. 246-250, 1969.

IANOVICI, V. - FLOREA, N.: (The accumulation of salts in the soils of quaternary plains in Rumania). (F,e) Studii Pedol. Ser. C., (14) 5-20, 1964.

IMRE, J.: (Classification of salt affected soils according to their quality and suitability for tree growth /Hungary/). (H,r,e) Kisérletügyi Közlem., 54A (3) 63-96, 1961.
(Increasing the fertility of solonetz soils /USSR/). (R) Akad. Nauk Ukr. SSR, Kiev, 1954.

IVANOVA, E.N.: (Tentative classification of steppe solonetzes /USSR/). (R,e) Pochvovedenie, (3) 14-26, 1963.

IVANOVA, E.N.: (Tentative classification of steppe solonetzes and solods /USSR/). (R,e) Pochvovedenie (4) 20-29, 1963.

IVANOVA, E.N.: (Solonetzes of the chestnut zone in the region between the rivers Volga and Ural /USSR/. (R) In: "Soils in the northern part of the Precaspian complex lowland and their ameliorative characteristics", 114-155. Nauka, Moskva, 1964.

IVANOVA, E.N. - ROZANOV, A.N.: (Classification of salt affected soils /USSR/). (R,e) Pochvovedenie, (7) 44-52, 1939.

IVANOVA, E.N. - ROZOV, N.N.: (Classification of soils in the steppe zones of USSR). (R) Pochvovedenie, (12) 48-59, 1958.

KELLEY, W.P.: The reclamation of alkali soils. (E) Reinhold, New York, 1937.

KELLEY, W.P.: Cation exchange in soils. (E) Reinhold, New York, 1948.

KELLEY, W.P.: Alkali soils. (E) Reinhold, New York, 1951.

KIZILOVA, A.A.: Movement of easily soluble salts in solonchak soils under leaching. (E.f) Arid Zone Res., 14 (Salinity problems in the arid zones. Proc. Teheran Symp.) 227-232, 1961.

KOLPAKOV, V.V.: (Effect of a water table rise on evaporation and the water soluble salt regime of soils in the Presivash region /USSR/). (R,e) Izv. timiryazev. sel'--khoz. Akad., (1) 132-147, 1961.

KONDORSKAYA, N.I.: (Geographical distribution of soda-affected soils in the Soviet Union). (R,e) Pochvovedenie, (9) 10-16, 1965.

KONDORSKAYA, N.I.: (Areas of present salt accumulation in the soils of the Soviet Union). (R,e) Pochvovedenie, (4) 44-55, 1967.

KOTIN, N.I.: (Calcareous solonetzes in the western part of the Ural plateau /USSR/). (R,e) Pochvovedenie, (7) 67-76, 1962.

KOVALISHIN, D.I.: (Genetic characteristics of solod soils in the Dnieper left side Ukrainian forest steppes /USSR/). (R) Agrokhim. Pochv., (3) 69-75, 1967.

KOVDA, V.A.: (Solonchaks and solonetzes /USSR/). (R) Izd. Akad. Nauk SSSR, Moskva, 1937.

KOVDA, V.A.: (Origin and regime of salt affected soils /USSR/). (R) I.-II.Izd.Akad. Nauk SSSR, Moskva, 1947.

KOVDA, V.A.: (Soils of the Prekaspian-bottomland. NW part /USSR/). (R) Izd. Akad. Nauk SSSR, Moskva, 1950.

KOVDA, V.A.: (Geochemistry of deserts in the USSR). (R) Izd. Akad. Nauk SSSR, Moskva, 1954.

KOVDA, V.A.: Principles of the theory and practice of reclamation and utilization of saline soils in the arid zones. (E,f) Arid Zone Res., 14 (Salinity problems in the arid zones. Proc. Teheran Symp.) 201-213, 1961.

KOVDA, V.A.: Alkaline soda-saline soils. (E,r) Agrokém. Talajt. 14 Suppl. 15-82. 1965.

KOVDA, V.A. - EGOROV, V.V. (Editors): (Drainage in the reclamation of salinized soils). (R) Izd. Akad. Nauk SSSR, Moskva, 1958. (E): National Found, Washington, D.C., US Dept. Agric., 1960.

KOVDA, V.A. - MINASHINA, N.G.: (Editors): (Irrigation and drainage of salt affected soils /USSR/). (R) Izd. Nauka, Moskva, 1967.

KOVDA, V.A. - SAMOILOVA, E.M.: Some problems of soda salinity. (E) Agrokém. Talajt. 18 Suppl. 21-36, 1969.

KOVDA, V.A. - ROZOV, N.N. - SAMBUR, G.N.: (Improvement and utilization of solonetzes /USSR/). (R) Izd. Akad. Nauk SSSR, Moskva, 1950.

KOZLOVSKY, F.I. - KORNBLYUM, E.A.: (Soil ameliorative conditions in the Volga- -Akhtuba bottomland /USSR/) in connection with its development and evolution).(R, e) Pochvovedenie, (7) 73-84, 1963.

KUNTZE, H.: (Formation, properties, utilization and reclamation of marsh soils). (G) Phosphorsäure, 20 250-276, 1960.

KUZMIN, V.A.: (Occurrence of salt affected soils under forest /USSR/). (R, e) Poch- vovedenie, (1) 111-114, 1962.

LATKOVICS, I.: (Effect of nitrogen containing materials on the efficiency of solonetz amelioration in the Hungarian Plain). (R, e) Agrokém. Talajt. 14 Suppl. 341-344, 1965.

LOPATNIK, J. - CERVENKA, L.: (Increasing the fertility of salt affected soils using gypsum and leaching /Czechoslovakia/). (Cz, r, g) Rostl. Vyroba, 34 575-590, 1961.

MAIANU, A.: (Secondary soil salinization /Rumania/). (Rm, e, r) Ed. Acad. RPR, Bu- charest, 1964.

MAIANU, A.: (Critical depth and critical mineralization degree of the ground-water in the lower Danube and Tisza plains /Rumania/ and factors affecting their varia- tions).(Rm, e, f, r) An. Sect. Pedol. 1964, 32 257-268, 1965.

MAIANU, A. - AKSENOVA, I.: (Accumulation of alkali carbonates and bicarbonates in mineralized ground-waters and in saline and alkali soils of Rumania). (Rm, e, f, r) An. Sect. Pedol. 1964, 32 267-282, 1965.

MAIANU, A. - AKSENOVA, L.: (Accumulation of alkali carbonates and bicarbonates in mineralized ground-waters and in saline and alkali soils of Rumania). (R, e) Ag- rokém. Talajt., 14 Suppl. 401-410, 1965.

MAMEDOV, R.: (Solonetz soils in Azerbaidzhan /USSR/). (R) Dokl. Akad. Nauk azer- baidzhan SSR, 23 (5), 1954.

MARBUT, C.F.: Soils of the United States. (E) USDA Atlas Amer. Agric., 1-98., 1936.

MAVROCORDAT, G. - NICOLAU, M. - ATANASIU, G.: (Experimental study of the improvement of salt affected soils /Rumania/). (Rm, r, e) Studii Cers. Agron., Cluj, 14 49-63, 1963.

MILJKOVIC, N.: (Characteristics of saline and alkali soils in Voyvodina/Yugoslavia/). (Sr-Cr, e) Izd. Saveza Vodn. Zajednica NRS, Novi Sad, 1963.

MILJKOVIC, N. - IVKOVIC, N. - GENADIC, J.: Chemical composition and season- al variation of (phreatic) ground-waters in Voyvodina /Yugoslavia/). (E) J.A.H. Con- gress, Hannover, 1965.

MILJKOVIC, N. - PLAMENAC, N.: Improvement of solonetz soils in Yugoslavia.(E) Agrokém. Talajt. 18 Suppl. 377-384, 1969.

MILJKOVIC, N. - PLAMENAC, N.: Solonetz soils of Yugoslavia, their properties and possibilities of utilization. (E) In: I. Szabolcs: "European solonetz soils and their reclamation". 151-164. Akadémiai Kiadó, Budapest, 1971.

MILJKOVIC, N. - EBERHARDT, Z.D. - AYERS, D.A.: Salt affected soils of Yugo- slavia. (E) Soil Sci. 88 51-55, 1959.

MOLEN, W.H. van der - BOUMANS, J.H.: Drainage requirements of irrigated soils in relation to salinity. (E,f,g) Trans. 8th Int. Congr. Soil Sci., 2 847-854, 1964.

MOZHEIKO, A.M.: (Formation of the solonetz horizon in the soils of the middle Prednieper region /USSR/). (R) Trudy kharkov. sel'khoz. Inst., 27 63-152, 1960.

MOZHEIKO, A.M.: Chemical reclamation of sodic solonetzes in the southern part of the middle Dnieper region /USSR/ by application of gypsum and calcium chloride. (E) Agrokém. Talajt. 18 Suppl. 310-314, 1969.

MUNTEANU, I. - IONESCU, M.: (Bog soils with maritime salinization in the Danube delta /Rumania/). (F,e,g) Trans. 8th Int. Congr. Soil Sci., 5 667-674, 1964.

MURAKÖZY, K.: (The soil). (H) Term. Tud. Közl., XXXIII (5) 593., 713., 1902.

MURGOCI, G.: (Soil survey and soil cartography). (F) Bucuresti Cartea Romin.,1924.

NEMES, M. et al.: (Conductometrical method for the determination of soil salinity and the classification of saline soils in Transylvania /Rumania/). (Rm,r,e) Lucr. st. Agron. "Dr. Petru Groza", Cluj, 19 27-34, 1963.

NEUGEBAUER, V. - MILJKOVIC, N.: General review of salt affected soils of Yugoslavia. (E) IRYDA: información. Salt Affected Soils Subcommission Meeting, Sevilla, May 1971, 191-198.

NIKITINA, A.I.: (Signs of sodic salinization of soil in Moldavia /USSR/). (R) Vopr. issled. ispol'z. pochv. Moldav., 3 43-58, 1965.

NIKODLJEVIC, V.: (Chemical and physical properties of a solonchak at Melenci and a solonetz at Kumane /Banat region, Yugoslavia/). (Sr-Cr,e) Arh. poljopr. Nauke, 16 114-123, 1963.

NOVIKOVA, A.V.: (Geochemical characteristics of salt accumulation processes and salt regime in the soils of the Crimean steppe /USSR/. The improvement of solonetzes and land utilization under irrigation). (R) Trudy kharkov. sel'khoz. Inst.,39 241-359, 1962.

NOVIKOVA, A.V.: (Prognosis of secondary salinization and alkalinization of soils due to irrigation /USSR/). (R) Agrokhim. Pochv., (5) 3-9, 1967.

OBREJANU, G. - SANDU, G.: Amelioration of solonetz and solonetzized soils in the Socialist Republic of Rumania. (E) In: I. Szabolcs: "European solonetz soils and their reclamation". 99-130. Akadémiai Kiadó, Budapest, 1971.

OBREJANU, G. - SANDU, G.: (Salinization and alkalinization of soils in the Danube plain /Rumania/). (Rm,r,f,e) Probl. Agric., 18 (5) 14-27, 1966.

OBREJANU, G. - MAIANU, A. - AKSENOVA, I.: (Salt accumulation in mineralized ground-waters and saline soils of floodplains in the lower Danube bottomland (Rumania/). (R,e) Pochvovedenie, (8) 44-62, 1964.

OBREJANU, G. - MAIANU, A. - ALBESCU, I.: (The prevention and control of secondary salinization of diked and irrigated soils in the lower Danube floodplain /Rumania/). (Rm) Stiinta Solului, 1 (1) 64-72, 1963.

OBREJANU, G. et al.: (Soils of the Danube floodplain /Rumania/ and problems of their salinization, solodization and amelioration). (R) Pochvovedenie (4) 55-66,1967.

OGANESIAN, K.A.: (Sodic soils in the Arazhdian steppe /USSR/). (R) Trudy Inst. pocsv. agrochem. Arm. SSR., 2, 1963.

OPREA, C.V. (Saline and alkali soils in Rumania and their amelioration). (Rm,r,f) Biol. stiint. Agric., 9 (1-2) 177-230, 1962.

OPREA, C.V. (The genesis and amelioration of salt affected soils in the western lowlands of Rumania). (R,e) Agrokém. Talajt. 14 Suppl. 183-188, 1965.

OPREA, C.V. - ANASTASESCU, I.: (Saline and alkali soils in the Eastern part of the Tisza Plain /Rumania/). (F) IRYDA: información. Salt Affected Soils Subcommission Meeting, Sevilla, May 1971, 77-86.

OPREA, C.V. - STEPANESCU, E. - VLAS, I.: Saline and alkali soils and their amelioration /Rumania/). (Rm) Ceres, Bucuresti, 1971.

OPREA, C.V. et al.: (Soils in the western part of Rumania, their nomenclature and classification). (Rm,r,f) Biol. stiint. Agric., 9 (1-2) 25-136, 1962.
(Origin of salt affected soils and their amelioration /USSR/).(R) Trudy pochv.Inst. Dokuchaeva, 44 1-414, Moskva, 1954.

ÖZTAN, B.: Saline and alkali soils in Turkey. (E) Semin. Land Class. and Soil Survey, Adana, Turkey, 61-70, 1960.

PAK, K.P.: Solonetzes of the European part of the USSR and their reclamation. In: I. Szabolcs: "European solonetz soils and their reclamation". 139-149. Akadémiai Kiadó, Budapest, 1971.

PAK, K.P. et al. (Solonetz amelioration methods in different zones of the Soviet Union). (E,f,g) Trans. 8th Int. Congr. Soil Sci., 2 891-896, 1964.

PANKOVA, E.I. et al. (Characteristics of the main regions of salt affected soils /USSR/). (R) In: Egorov, V.V. - Bazilevich, N.I. "Salt affected soils in the European part of USSR and Transcaucasian Soviet Republics". 204-267. Nauchn. Trudy pochv. Inst. Dokuchaeva, Moskva, 1973.

PANOV, N.P. - NERETIN, G.I.: (Solonetz complexes in the north-eastern Precaucasus region /USSR/). (R,e) Izv. timiryazev. sel'khoz. Akad., (5) 149-156, 1968.

PAXINOS, S.A. - ALEXIADES, K.A.: (Effect of the synthetic soil conditioner /Rohagit S 7687/ on the structure of nonsaline alkali soils /Greece/). (Gr,e) Aristot. Panepist. Thessalonikis Epet. Geopon. Dasolog. Skol., 31-51, 1961.

PEKATOROS, L.G.: (Salt accumulation processes in the soils of the southern Bug floodplain /USSR/). (R,e) Pochvovedenie, (8) 29-36, 1960.

PEKATOROS, L.G.: (Secondary salinization of soils in the Dnieper left bank bottomlands and Danube delta, in the Ukraine). (R,e) Pochvovedenie, (2) 26-36, 1962.

PEKATOROS, L.G.: (Salinity and alkalinity of floodplain soils in the western part of the Black Sea lowland /USSR/). (R,e) Pochvovedenie, (11) 56-66, 1967.

PEKATOROS, L.G.: (Secondary salinization of river plain soils in the steppe zone of the right bank Ukraine). (R,e) Pochvovedenie, (8) 54-65, 1969.

PERSHINA, M.N.: (Classification and characteristics of solonchaks /USSR/).(R) Dokl. mosk. sel'khoz.Akad. K.A. Timiryazeva, 89 134-139, 1963.

PERSHINA, M.N. - DODOLINA, V.T.: (Short classification and diagnostics of solonetzes /USSR/). (R) Dokl. mosk. selkhoz. Akad. K.A. Timiryazeva, 84 46-51, 1963.

PLAMENAC. N. - MILJKOVIC, M.: Contribution to the programme of the field experiment on the reclamation of solonetz soils in Yugoslavia. (E) IRYDA: información. Salt Affected Soils Subcommission Meeting, Sevilla, May 1971, 63-66.

PODYMOV, V.P. - SULIN, I.V.: (Solonchaks in the Dniester floodplain /USSR/). (R) Vopr. issled. ispol's. pochv. Moldavii, (e) 59-68, 1965.

POLYNOV, B.B.: (Selected papers). (R) Izd. Akad. Nauk SSSR, Moskva, 1956.

POPOV, A.A.: (Salt regime of soils in the Volga-Akhtuba bottomland /USSR/ in connection with controlling the flow rate of the Volga river by the Volgograd reservoir). (R, e) Pochvovedenie, (5) 57-68, 1964.

PRETTENHOFFER, I.: (Amelioration of non-calcareous alkali soils /meadow solonetzes/ by subsoil loosening in Hungary). (G,e,f,r) Agrokém. Talajt. 13 Suppl.227--235, 1964.

PRETTENHOFFER, I.: Amelioration of sodic solonetz soils in the region east of the river Tisza /Hungary/). (E,r) Agrokém. Talajt. 14 Suppl. 323-328, 1965.

PRETTENHOFFER, I.: (Utilization and amelioration of Hungarian salt affected soils in the Transtisza region). (H) Akadémiai Kiadó, Budapest, 1969.

(Problems of salinization and alkalinization of soils and water sources /USSR/). (R) Izd. Akad. Nauk SSSR, Moskva, 1960.

RADANOVIC, R. - PLAMENAC, N. - MILJKOVIC, N.: Appearance and some characteristics of potential salt affected soils in southern part of Pannonian Plain in Yugoslavia. (E) IRYDA: información. Salt Affected Soils Subcommission Meeting, Sevilla, May 1971, 49-56.

RAIKOV, L.: Reclamation of solonetz soils in Bulgaria. (E) In: I. Szabolcs: "European solonetz soils and their reclamation". 35-47. Akadémiai Kiadó, Budapest, 1971.

RAIKOV, L. - BEHAR, A.: (Salt affected soils at Belozem, near Plovdiv /Bulgaria/). (B,r,e) Izv. tsentr. nauchno-issled. Inst. Pochv. Agrotech. "N. Pushkarov",3 25-58, 1962.

RAIKOV, L. - KAVARDZHIEV, YA.: (Amelioration of sodic solonetzes in Bulgaria). (R,e) Agrokém. Talajt. 14 Suppl. 225-228, 1965.

RAIKOV, L. - KAVARDZHIEV, YA. - STAIKOV, S.: (Raising the fertility of meadow solonetz soils /Bulgaria/). (B,r,e) Nach. Ses. povysh. plod. pochv. Bolgarii, 191--198, 1965.

RAIKOV. L. - KAVARDZHIEV, YA. - VARBANOVA, Z.: Residual effects of chemical treatment and tillage operations for improvement of sodic solonetz soils. (E) IRYDA: información. Salt Affected Soils Subcommission Meeting, Sevilla, May 1971, 39-48.

RAIKOV, L. et al.: (Amelioration of salt affected soils /Bulgaria/). (B,r,e) Izd.Bolg. Akad. Nauk, Sofia, 1966.

RAIKOV, V.: (Microflora of Bulgarian salt affected soils. (R) Pochvovedenie, (4) 104-116, 1965.

RICHARDS, L.A.: Diagnosis and improvement of saline and alkali soils. (E) US Dept. Agric. Handbook No. 60., 1954.

RODE, A.A.: A biogeocoenological approach to the solution of land reclamation problems. (E) Agrokém. Talajt. 18 Suppl. 278-282, 1969.

ROQUERO, C.: (Saline soils in Spain). (Sp) IRYDA: información. Salt Affected Soils Subcommission Meeting, Sevilla, May 1971, 199-202.

ROTINI, O.T.: (Degradation and reclaiming of clay soils in Tuscany /Italy/). (I) Italia agric. 105 319-331, 1968.

ROZANOV, A.N.: System of scientific investigations in projects for the irrigation and reclamation of saline soils. (E, f) Arid Zone Res., 14 (Salinity problems in the arid zones. Proc. Teheran Symp.) 223-226, 1961.

Salinity problems in the arid zones. (E, f) Proc. Teheran Symp. Arid Zone Res., 14 1-395, UNESCO, Paris, 1961.

SAMBUR, G.N.: (Solonetz soils of the Ukraine, their genesis, agricultural character-istics and improvement). (R) Gossel'khozgiz UkrSSR, Kiev, 1962.

SANDU, G.: (Salt sources of the salt accumulation and alkalinization processes in the soils of the Danube valley /Rumania/). (Rm, e, f, r) An. Sect. Pedol. 1966. 34 333--345, 1967.

SANDU, G.: (Salt regime of saline and alkali soils after amelioration in the Romanian Plain). (F) IRYDA: información. Salt Affected Soils Subcommission Meeting, Sevilla, May 1971, 113-142.

SANDU, G. et al.: (Amelioration of sodic soils by gypsum and fertilizers /Socodor Experimental Centre, Rumania/). (R, e) Agrokém. Talajt. 14 Suppl. 211-224, 1965.

SANDU, G. et al.: (Study of ground-water mineralization and the hydrosaline regime of soils in the Calarasi terrace irrigation systems /Rumania/). (Rm, e, f, r) An.Sect. Pedol. 1965, 33 323-343, 1966.

SEDLÁK, S. - CERVENKA, L.: (Salt affected soils in the Malcice loess plateau of the East Slovakian Plain /Czechoslovakia/). (Cz, r, e) Vedecke práce Labor.Podozn. Bratislava, 2 1967.

SERVANT, J.: (Main characteristics of saline and alkali soils in the Roussillon Plain /France/). (F) Bull. Ass. Fr. Etude du Sol, No.3. 1969.

SERVANT, J. - FAVROT, J.C.: (Saline soils of the Languedoc-Roussilon coast /France/. Proposed classification). (F) Trans. Conf. Mediterranean Soils, Madrid, 1966, 1-6, 1967.

SERVANT, J. - SERVAT, E.: (Introduction to the study of saline soils on the Lan-guedoc-Roussillon coastline /France/). (F, e, g, r) Ann. Agron., 17 53-73, 1966.

SERVANT, J. - SERVAT, E.: (Sodic soils of France. General characteristics and criteria of classification). (F) IRYDA: información. Salt Affected Soils Subcommis-sion Meeting, Sevilla, May 1971, 165-182.

SHOPSKI, N.: (Salinization and desalinization of soils and ground-waters in the Kara-boaz lowland /Bulgaria/). (B, r, e) Pochvozn. Agrokh., 1 (5) 413-423, 1966.

'SIGMOND, A.A.:Hungarian alkali soils and methods of their reclamation. (E) Printing Office, Berkeley Calif. Univ., 1-156, 1927.

SINGH, H.P.: Humus and nutrient status of salt affected soils in relation to their gen-esis, with special reference to the soils of the Hungarian Plain. (E) Agrokém.Ta-lajt. 17 Suppl. 73-89, 1968.

SIPOS, S. - BOCSKAI, J.: (Efficiency of liming on a meadow solonetz soil turning into steppe formation in the case of various agrotechnical factors /Hungary/). (H, e, sp, r) Agrokém. Talajt. 15 491-506, 1966.

SIPOS, S. - BOCSKAI, J.: (Increasing the fertility of solonetz soils by differential improvement of genetic soil horizons /Hungary/). (H, r, e, g) Talajtermékenység 3 (2) 91-114, 1968.

SKURTUL, A.G.: (Characteristics of salt affected soils in the small river floodplains of southern Moldavia /USSR/). (R) Trudy Moldavsk. nauchno-issled. Inst. oroshaem. zemled. ovoshch., 3 55-66, 1961.

SLAVNII, YU. A. - KAURICHEVA, Z.N.: (Characteristics of soil salinization on the Prevolga ridge /USSR/). (R, e) Pochvovedenie (5) 121-130, 1967.

SMITH, G.D.: Soil classification. A comprehensive system. 7th Approximation. (E) Soil Survey Staff, Soil Conserv. USDA, 1960.

(Soil amelioration in the Kura-Araks Lowland /USSR/). (R) Izd. Akad. Nauk SSSR, Moskva, 1962.

(Soil amelioration studies in the Volga-Akhtuba bottomland and Volga deltas /USSR/). (R) Izd. Nauka Moskva, 1958.

(Soil ameliorative and ecological conditions of the Ergenya region and the northwestern part of the Precaspian lowland /USSR/). Izd. Akad. Nauk SSSR, Moskva, 1961.

(Soils in the northern part of the Precaspian complex lowland /USSR/ and their ameliorative characteristics). (R) Izd. Nauka, Moskva, 1964.

SOKOLOVSKY, S.P.: (Characteristics of recent soil salinization in the delta of the Terek-river /USSR/). (R, e) Pochvovedenie, (5) 72-81, 1960.

SOMMERKAMP, G. - GALENSA, F. - KUNTZE, H.: (German marshes. A review of the present state of research). (G, e) Z. Kulturtechnik, 1 257-288, 1960.

STEBUTT, A.: (General soil science). (G) Borntraeger, Berlin, 1930.

STEPANESCU, E.: (Main factors affecting the development of saline soils in the Cris plain /Rumania/). (Rm, e, f, r) An. Sect. Pedol. 1964, 32 339-349, 1965.

STORCH, G.· (The influence of land reclamation /drainage, levelling of beds and liming/ on the physical properties and water content of a brakish marsh soil, rich in silt /Germany/). (G, e) Z. Kulturtechnik, 2 297-314, 1961.

SZABOLCS, I.: (The soils of Hortobágy region /Hungary/). (H, r) Mezőgazd. Kiadó, Budapest, 1954.

SZABOLCS, I.: Degradation of irrigated soils in Hungary. (E, g, f) Trans. 7th Int. Congr. Soil Sci., 1 638-644, 1960.

SZABOLCS, I.: (Effect of water regulations and irrigation on soil formation processes in the region beyond the river Tisza /Hungary/). (H) Akadémiai Kiadó, Budapest, 1961.

SZABOLCS, I.: The influence of irrigation water of high sodium carbonate content on soils /Hungary/). (E, f, g, r) Agrokém. Talajt. 13 Suppl. 237-246, 1964.

SZABOLCS, I.: Salt affected soils in Hungary. (E, r) Agrokém. Talajt. 14 Suppl. 275-306, 1965.

SZABOLCS, I.: Recent problems of investigating and utilizing salt affected soils. (E, r,g,f,sp) Beitr. trop. Landw. Vet. Med., 4 5-15, 1966.

SZABOLCS, I.: The influence of sodium carbonate on soil forming processes and on soil properties /Hungary/). (E) Agrokém. Talajt. 18 Suppl. 37-68, 1968.

SZABOLCS, I.: Genetics, geography and properties of European solonetz soils. (E) Vedecke Práce Lab. Podoznalectva, Bratislava, 3 193-204, 1970.

SZABOLCS, I. (Ed.): European solonetz soils and their reclamation. (E) Akadémiai Kiadó, Budapest, 1971.

SZABOLCS, I.: Solonetz soils in Europe. Their formation and properties with particular regard to utilization. (E) In: I. Szabolcs (Ed.): "European solonetz soils and their reclamation". 9-34. Akadémiai Kiadó, Budapest, 1971.

SZABOLCS, I.: (European salt affected soils and their utilization). (H,g) Földrajzi Közlem., XIX/XCV. 145-154, 1971.

SZABOLCS, I.: (Soda affected soils and solonetzes). (H) Agrokém Talajt., 21 415--434, 1972.

SZABOLCS, I. - CERVENKA, L.: (Characteristics of salt affected soils in Slovakia). (Cz,e,r) Vedecke práce Labor. Podozn. Bratislava, 3 211-227, 1968.

SZABOLCS, I. - DARAB, K.: (Dynamics of soluble salts in irrigated soils /Hungary/). (H,r,e) Agrokém. Talajt. 4 251-264, 1955.

SZABOLCS, I. - DARAB, K. Accumulation and dynamism of silicic acid in irrigated alkali /"szik"/ soils. (E) Acta agron. Hung., 8 213-235, 1958.

SZABOLCS, I. - DARAB, K.: Investigation of the effect of anions on Na-Ca ion exchange. (E) Agrokém. Talajt. 17 Suppl. 21-40, 1968.

SZABOLCS, I. - DARAB, K.: Salt balance and salt transport processes in irrigated soils /Hungary/). (E,f,g) Trans. 9th Int. Congr. Soil Sci., 1 491-498, 1968.

SZABOLCS, I. - JASSÓ, F.: (The classification of Hungarian salt affected soils). (H, r,g) Agrokém. Talajt. 8 281-300. 1959.

SZABOLCS, I. - JASSÓ, F.: (Genetical types and regularities in the occurrence of salt affected soils in the region between the rivers Danube and Tisza /Hungary/). (H,r,e) Agrokém. Talajt. 10 173-190, 1961.

SZABOLCS, I. - LESZTÁK, V.: The movements of different salt solutions in soil profiless (E) "Water in the unsaturated zone". Symp. Wageningen. Publ. No. 83. AIHS. 611-621. IASH/AISH-UNESCO, Paris, 1969.

SZABOLCS, I. - MÁTÉ, F.: (Genetics of salt affected soils in the Hortobágy region /Hungary/). (H,r,f) Agrokém. Talajt. 4 31-38, 1955.

SZABOLCS, I. - DARAB, K. - VÁRALLYAY, G.: Salt balance of irrigated soils /Hungary/). (E,r,g,f,sp) Beitr. trop. Landw. Vet. Med., 4 (2) 123-135, 1966.

SZABOLCS, I. - DARAB, K. - VÁRALLYAY, G.: (Salt balances for the prediction and prevention of secondary salinization and alkalinization of irrigated soils /Hungary/) (Alb, b,cz,g,h,mong,p,r,rm) Nemzetk. Mezőgazd. Szemle, 13 (5) 46-50, 1969.

SZABOLCS, I. - DARAB, K. - VÁRALLYAY, G.: Methods for the prognosis of salinization and alkalinization due to irrigation in the Hungarian plain. (E) Agrokém.Talajt. 18 Suppl. 351-376, 1969.

61

SZABOLCS, I. - DARAB, K. - VÁRALLYAY, G.: (Prediction and prevention of secondary salinization and alkalinization processes due to irrigation in the Hungarian Plain). (R, e) Pochvovedenie, (1) 115-124, 1972.

SZABOLCS, I. et al.: (Soil alkalinisation studies in model experiments). (H, e, r) Agrokém. Talajt. 5 297-306, 1956.

SZABOLCS, I. et al.: The types of salt affected soils in Hungary and their utilization with particular regard to irrigation and leaching. (E, f) Trans. 6th Congr. I.C.I.D., New-Delhi, Q19 R14 171-186, 1966.

SZENDREI, G.: (Micromorphological examination of salt affected soils in the region of Kiskunság, Hungary). (H, e, f, r) Agrokém. Talajt. 19 231-242, 1970.

TALSMA, T.: Control of saline ground-water. (E, dutch) Meded. Landbouwhogeschool, Wageningen, 63 (10) 1-68, 1963.

TEIMUROV, K.G.: (Leaching of sulphate solonchaks with the preapplication of chemicals /USSR/). (R) Khlopkovodstvo, 14 44-45, 1964.

TEIMUROV, K.G.: (Leaching of heavy textured saline soils on the Kura-Araks lowland /Azerbaidzhan SSR, Soviet Union/ using chemical amendments). (H, e, g, r) Agrokém. Talajt. 21 293-314, 1972.

TEIXEIRA, A.J.S.: (Thiosols /Portugal/). (Port.) Pedologia, 2 7-13, 1967.

TEIXEIRA, A.J.S. - VEIGUINHA, A.A. - ALVIM, A.J.S.: Reclamation and use of marine saline soils /Portugal/. (E, f, g) Trans. 8th Int. Congr. Soil Sci., 2 931--935, 1964.

TÓTH, B. et al.: (Afforestation of salt affected soils /Hungary/). (H) Akadémiai Kiadó, Budapest, 1972.

TREITZ, P.: (Saline and alkali soils /Hungary/). (H) Pátria, Budapest, 1924.

VADYUNINA, A.F.: Meliorative effect of direct electric current on leaching solonchakous solonetzes /USSR/. (E, f, g) Trans. 9th Int. Congr. Soil Sci., 1 455-463, 1968.

VÁRALLYAY, G.: (A peculiar case of sodic alkalinization in Hungary). (R, e) Agrokém. Talajt. 14 Suppl. 33-340. 1965.

VÁRALLYAY, G.: (Salt balances of soils in the region between the rivers Danube and Tisza /Hungary/. I. Salt balances under natural /non-irrigated/ conditions). (H, e, f, r) Agrokém. Talajt. 15 423-452, 1966.

VÁRALLYAY, G.: (Salt balances of soils in the regions between the rivers Danube and Tisza /Hungary/. II. Salt balances under irrigated conditions). (H, e, f, r) Agrokém. Talajt. 16 27-56, 1967.

VÁRALLYAY, G.: (Salt accumulation processes in the soils of the Hungarian Danube Valley). (H, e, r, f) Agrokém. Talajt. 16 327-356, 1967.

VÁRALLYAY, G.: Salt accumulation processes in the Hungarian Danube Valley. (E, f, g) Trans. 9th Int. Congr. Soil Sci., 1 371-380, 1968.

VÁRALLYAY, G.: (Hydraulic conductivity of salt affected soils in the Hungarian Plain). (H, e, f, r) Agrokém. Talajt. 21 57-88, 1972.

VÁRALLYAY, G.: Application of the unsaturated flow theory in the prognosis of salinization from the groundwater. (E) ICID 9th European Conf., Q2 R2.1/4. 1-11, Budapest, 1973.

VÁRALLYAY, G. - SZABOLCS, I.: (Salt affected soils in Transdanubia /Hungary/. III. Salt affected soils in the Mezőkövesd region). (H, e, f, r) Agrokém. Talajt. 15 1-42, 1966.

VERHOEVEN, B.: Leaching of sodic soils as influenced by application of gypsum. (E, r) Agrokém. Talajt. 14 Suppl. 263-268, 1965.

VERNANDER, N.V. et al.: (Soils of Ukrainian SSR). (R) Sel'khozizdat, Kiev-Khar'-kov, 1951.

VILENSKY, D.G.: (Salt affected soils, their origin, composition and methods of amelioration /USSR/). (R) Novaya derevnya, Moskva, 1924.

VIL'IAMS, V.R.: (Selected papers). (R) Izd. Akad. Nauk SSSR, Moskva, 1950.

VILLAR, E.H. del: (Soils of the Iberian peninsula). (Sp) Pedidos en Espana, Madrid, 1937.

VODENICHAROV, I.: (Salt affected soils in the Plovdiv district /Bulgaria/). (B, r, e) Izd. Bolg. Akad. Nauk, Sofia, 1968.

VOLOBUEV, V.R.: (Salt affected soils in Azerbaidzhan /USSR/, their formation and amelioration). (R) Izd. Akad. Nauk Azerbaidzhan SSR, Baku, 1948.

VOLOBUEV, V.R.: (Genetic types of salt affected soils in the Kura-Araks lowland /USSR/). (R) Izd. Akad. Nauk Azerbaidzhan SSR, Baku, 1965.

ZIMOVETS, B.A.: (Alteration of the state of reclamation land by irrigation in the southern Transvolga region /USSR/). (R, e) Pochvovedenie, (6) 76-85, 1968.

ZIVKOVIC, B.: (Salinization and comparative characteristics of normal soils, soils in the process of salinization and solonchaks in Vojvodina /Yugoslavia/). (Sr-Cr, e) Savremena poljoprivreda, 5 (Spec. ed.) 1-91. 1965.

— . —

If a paper was published in English, the original title is given. If a paper was published in other language, a short descriptive English translation of the original title is given in brackets.

The country in or after the title indicates that the author deals with the problems of salt affected soils occurring in that country. Lack of the indication of the country means that the author deals with general problems of salt affected soils.

A capital letter in square brackets following the title denotes the language in which the paper was written. A small letter denotes that the original paper contained a summary in another language, e.g. (H, r, g, e) - in Hungarian, with Russian, German and English summaries. The abbreviations of languages used in this bibliography are as follows: Alb. - Albanian, B - Bulgarian, Cz - Czechoslovakian, D - Danish, E - English, F - French, Finn. - Finnish, G - German, Gr - Greek, H - Hungarian, I - Italian, Mong. - Mongolian, P - Polish, Port. - Portugese, R - Russian, Rm - Romanian, Sp - Spanish, Sr-Cr - Serbo-Croatian, Ukr - Ukranian.

The titles and other abbreviations used in this bibliography are similar to those used by the Commonwealth Bureau of Soils (Harpenden, England) and are given in their list "Titles and Abbreviations of Publications Noted in Soils and Fertilizers 1966-67".

Kiadásért felel: Szabolcs István igazgató
Példányszám: 2100
Ivterjedelem: 5,6 A/5 ív
74267 DATORG F.v.: Harkai József
A két térképmelléklet a Kartográfia nyomdában készült

Additional material from *Salt Affected Soils in Europe*
ISBN 978-94-011-8638-4, is available at http://extras.springer.com